柏林联建住宅

Co-housing in Berlin

龚喆　李振宇　[德]菲利普·米塞尔维茨 著
by GONG Zhe　LI Zhenyu　Philipp Misselwitz

中国建筑工业出版社
CHINA ARCHITECTURE & BUILDING PRESS

图书在版编目(CIP)数据

柏林联建住宅 ：汉英对照 / 龚喆，李振宇，(德)米塞尔维茨著．-- 北京 ：中国建筑工业出版社，2015.12
ISBN 978-7-112-18795-9

Ⅰ．①柏… Ⅱ．①龚… ②李… ③米… Ⅲ．①住宅－建筑设计－柏林－汉、英 Ⅳ．①TU241

中国版本图书馆CIP数据核字(2015)第289436号

责任编辑：滕云飞　朱笑黎
责任译校：王梓笛

柏林联建住宅
Co-housing in Berlin
龚喆 李振宇 [德]菲利普·米塞尔维茨 著
*
中国建筑工业出版社出版、发行（北京西郊百万庄）
各地新华书店、建筑书店经销
上海盛通时代印刷有限公司制版
上海盛通时代印刷有限公司印刷
*
开本：889×1194毫米 1/20 印张：8⅕ 字数：246千字
2016年1月第一版 2016年1月第一次印刷
定价：58.00元
ISBN 978-7-112-18795-9
(28074)
版权所有 翻印必究
如有印装质量问题，可寄本社退换
（邮政编码 100037）

前言 | Preface

1

人因宅而立，宅因人得存。作为与人最密切的建筑形式，住宅建筑伴随时代的变迁发展出了众多的类型。现代住宅建筑按照建设主体划分大致有三类：其一是由权力机构所主导的针对指定住户的福利性住房；其二是由商人所主导的面向市场的商品性住房；其三是由居住者主导建设的自用性住房。

第一次世界大战之后，随着居住需求的变化，一种介乎三者之间的新的住宅类型“联建住宅”在欧洲悄然出现。这种住宅是由多个用户通过合作协商的方式联合兴建的。在一定程度上，它很好地结合了三类住宅的优点而又规避了各自的缺点，既能回应住户不同的个性化需求，又能保持相应的共性。惜乎联建住宅从组织到实施，总不如传统的三种方式那样富有效率：既不像政府那样集中政策资源，又不像企业那样可以运用好“看不见的手”进行专业化运作，也不像一户人家那样价值取向明确，又无需过于追求速度。

到了 21 世纪，信息化的技术方法，全球化的多样视野，个性化的居住需求，城市化的交往方式，为联建住宅提供了新的生长环境。在德国，“忽如一夜春风来，千树万树梨花开”，联建住宅迅速成为一种时尚。这是一种协作性的、以社区生活为导向的住房类型。在项目实践过程中，居民能够全过程参与住宅设计和项目管理。面对新时代住户多样化需求的局面，联建住房项目做出了积极回应。在柏林，通过不断地探索，联建住宅项目逐渐发展出了一套完整的运作机制，并形成了一大批摆脱传统居住模式，极富个性的住宅设计作品。

本书对 21 世纪以来柏林范围内联建住宅项目的类型与设计特征进行研究，探讨在有各方参与（包括项目团体成员）的情况下实现建筑设计的方法，以期在公共与私密，集体与个人，多元与统一等需求矛盾中，找出可供当代住宅设计借鉴的积极模式。

I

Housing is the closest architecture type to human beings with many different types. At all times and in all over the world housing can be classified into 3 types according to the main body of construction: welfare housing conducted by governments, commercial housing conducted by real estate developers and self-built housing by residents.

After WWI with the change of living requirements, a new type of housing sharing the features of the above three came into being in Europe. It is the so-called “Co-housing”, which provides a way for residents to co-built housing together. It well combines the advantages of the three and evades the disadvantages to some degree. However, Co-housing at that time was less effective than the main three types of housing: neither could Co-housing make full use of policies like governments, nor could it run professionally with the "invisible hand" like enterprises or with a clear purpose but without time limit like one family.

When it comes to 21st century, informational techniques, global perspectives, individualized living requirements and urbanized communication ways help built new growing environment for Co-housing. Co-housing, which is a type of collaborative housing and in which residents actively participate in the design and operation process of their own communities, has been favored increasingly in Germany. In Berlin, with continuous exploration, Co-housing has gradually developed a sound set of practical mechanisms and a large number of highly individualized residential design-works boomed getting rid of the conventional model.

Focusing on Co-housing’s typologies and design features in Berlin since 2000, this book explores positive and practical

本书选取了柏林 40 个住宅区以及单体加以研究，并对其中 19 个案例进行深入分析，配以相应的照片、图纸和数据资料，展现了柏林联建住宅的整体面貌，重点分析在城市设计以及建筑设计多样化、个性化探索中所采取的方法和已经取得的成绩。通过分析，期待对中国住宅建筑的多样化有相应的启发，对步履蹒跚的中国合作式建房提供一定的借鉴。

2

本书的诞生，首先是“同济大学——柏林工大城市设计双学位硕士研究生项目”的一项成果。2006 年，经过数年准备，由中国政府和德国政府资助的“共同学习、共同研究” 研究生联合项目正式开班，同济大学和柏林工大共同设定了两年四模块研究生课程，采取全英语教学，双方研究生在一起学习两年，第一年在柏林工大，第二年在同济大学。开始的几届由中国国家留学基金管理委员会（Chinese Scholarships Council）和德国德意志学术交流中心（Deutscher Akademischer Austausch Dienst）向学生提供奖学金，其后逐步过渡到常态化，虽不再有丰厚的奖学金，但一直不额外收取学费。该项目德中双方分别由 Peter Herrle 教授和李振宇担任创始项目主任。屈指算来，到今天已经有 1 0 届学生参加，受益的中德学生总计达到 200 人。

中德双方学生的研究论文，选题大多涉及到对方国家的城市建筑问题。这积极促进了彼此的了解和认识，拓宽了双方学生的视野，使他们认识到城市和建筑问题的参差多态。

龚喆是同济大学建筑与城市规划学院 2011 级硕士生，中方导师为李振宇，德方导师为菲利普·米塞尔维茨(Philipp Misselwitz）。龚喆 2012 年9 月到德国学习，在双方导师指导下选定柏林联建住宅为论文题目，由米塞尔维茨指导开

ways to design housing with the participation of multiple parties in order to find a new model for contemporary residential design resolving conflicts between the public and the private, the collective and the individual, the unified and the personalized.

The book selected 40 Co-housing projects including living communities and individual buildings. Based on the 40 cases, 19 of them are well displayed in detail with a lot of photos, drawings, and relevant data. Together they show the accomplishment of Co-housing in Berlin achieved in the pursuit of diversity and individuation both in urban design and architecture design.Through analysis, we are looking for inspiration and references to housing in China.

II

This book is firstly the achievement of Urban Design Master Dual Degree Program between Tongji University and TU Berlin. After several years of preparation the joint graduate program sponsored by Chinese government and German government started in 2006. This two-year program contains four learning modules via English teaching, which spend the first year in TU Berlin and the second year in Tongji University. In early years Chinese Scholarships Council and Deutscher Akademischer Austausch Dienst offered the participants scholarship and nowadays no additional tuition is charged. This program was established by Peter Herrle from TU Berlin and LI Zhenyu from Tongji University. Up to date it is in its 10th year and more than 200 students have taken part in this program.

The master dissertations in this program mainly deal with the

展调查工作；2013 年 9 月回到同济，由李振宇指导调整毕业论文并于 2014 年 6 月完成答辩。答辩中，中德双方教授和答辩委员给予很好的评价。本书就是在龚喆的中英双语硕士论文《柏林联建住宅类型与设计特征研究》的基础上修改完善而成的，这也是同济大学柏林工业大学城市设计双学位联合培养项目学生中，第一次正式双语出版的硕士论文。

本书的主要研究和写作工作由龚喆完成。李振宇帮助选择了研究课题，指导了研究提纲的设定和正文的修改，确定了本书的体例和形式；米塞尔维茨指导了柏林的现场调研，提供有关文献资料，协助进行了英文修改。

希望本书能为当今住宅建筑的设计和研究提供不同的思路，给出有价值的参考。

龚喆
李振宇
菲利普·米塞尔维茨

2015 年 12 月

urban and architectural issues in the opposite country which positively help the two countries know each other more deeply and help students from the two countries broaden their horizons and understand the diversity of issues in cities and architectures.

GONG Zhe started to study for his master degree in College of Architecture and Urban Planning in Tongji University in 2011. His Chinese supervisor is LI Zhenyu and German supervisor is Philipp Misselwitz. GONG Zhe went to Germany in September 2012 and chosen Co-housing in Berlin as his topic of master dissertation under the guidance of the two supervisors. With the help of Philipp Misselwitz, GONG Zhe started his field trips and research work in Berlin. When back to Tongji in September 2013, GONG Zhe finished his dissertation guided by LI Zhenyu. In June 2014 He finished his final defense and received good evaluation from Thesis Supervisory Committee. The book is based on GONG Zhe's bilingual master dissertation "Research on Co-housing's Typologies and Design Features in Berlin". This book is also the first published bilingual master dissertation of this program.

The main research and writing of this book was finished by GONG Zhe. LI Zhenyu helped choose the research topic, define the research outline and modified the main body of the book. Philipp Misselwitz guided the field trips in Berlin, provided relevant materials and modified the english version of the book. Wish this book offer different thought and valuable reference to the current housing design and research.

GONG Zhe
LI Zhenyu
Philipp Misselwitz
December 2015

目录 | Content

3 柏林联建住宅的设计特征 Design Features of Co-housing in Berlin

Part 1 Background Introduction 背景介绍

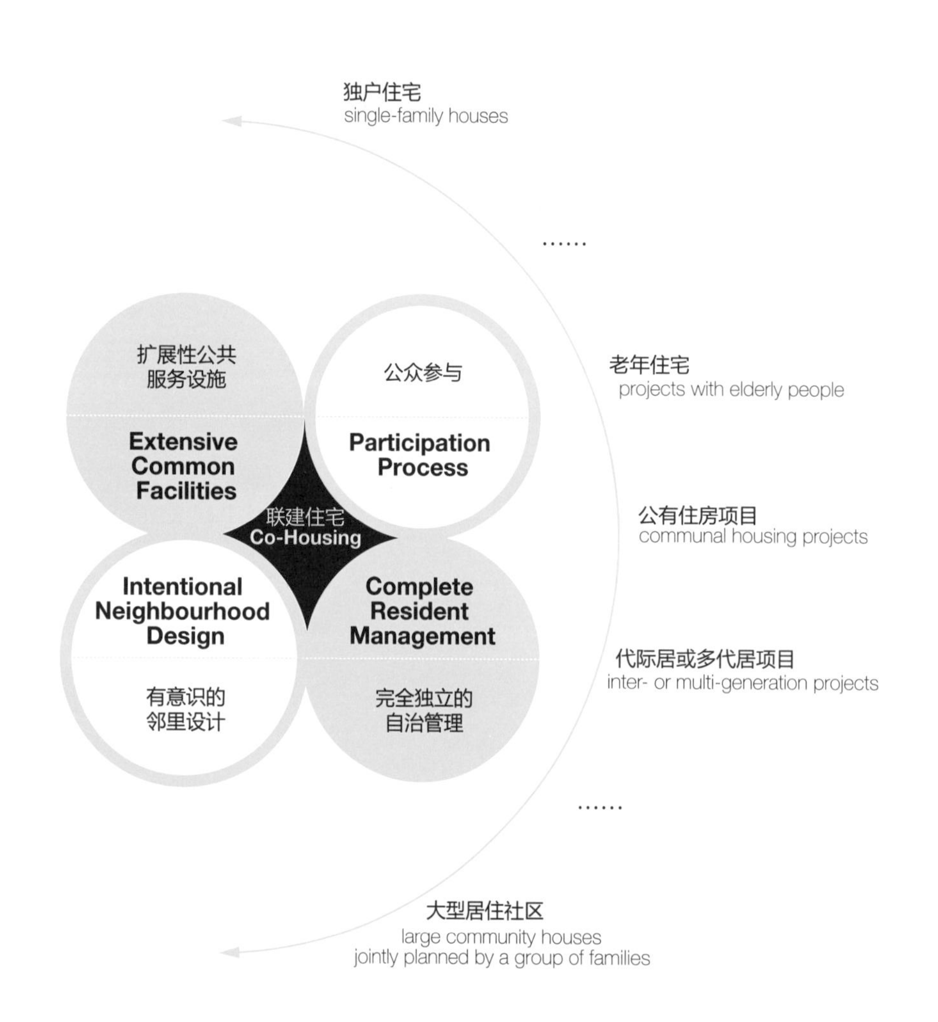

Fig.1.1 联建住宅的特征 Co-housing's Characteristics

1.1 联建住宅的相关定义

联建住宅是协作性住宅，居民积极参与住宅设计和实践过程[1]，是协作型、合作型、集体型和邻里导向型的住房项目(LaFond[2], 2011), 它拥有以下四大特性（Fig.1.1）：

1. 居民参与规划设计过程，是项目决策者；
2. 项目设计以邻里生活为导向，社区生活是日常生活的重要组成部分；
3. 项目内公共服务设施（厨房、餐厅）功能得到扩展，其空间占有率增大；
4. 居民以某种联合形式实施对住宅完全独立的管理。

联建住宅具有高度灵活性，这不仅体现在其规模可从小型独户住宅浮动到大型居住社区，也体现在其满足各类人群需求的、丰富多样的住宅空间形式。

联建住宅项目在实施过程中涉及到的个人、组织或者机构被称为其利益相关者。Experimentdays 12研讨会[3]将其分成了三大类：公共部门、市民团体（非政府和非营利组织）和私营部门。公共部门是指政府机构及其附属部门，例如由政府资助的代理或基层部门；市民团体是指由非政府人员发展的非营利性组织；私营部门则是营利性的公司。Fig.1.2展示了利益相关方的代表及其扮演的角色。

1 定义来源于 http://www.cohousing.org

2 Michael LaFond, 柏林联建住宅等自建项目专家，id22（Institute for Creative Sustainability）柏林可持续性创意研究所负责人，曾就职于柏林工业大学。

3 Experimentdays 每年由 id22（柏林可持续性创意研究所）举办，为联建住宅项目以及其他创意可持续性项目提供交流平台，以期将城市理解为经过设计的生活空间。自2003年以来，Experimentdays 为人们了解各式自组织住宅项目，寻找支持者以及思考未来城市发展方面提供了很多机会。同时它也为日常生活，艺术和政治等方面的探索和实践提供了场地。2012 年举办的这一届名为 Experimentdays 12。

1.1 Definition of Co-housing

Co-housing is a type of collaborative housing in which residents actively participate in the design and operation of their own neighborhoods[1]. And it is understood as collaborative, cooperatives, collective and community-oriented housing project (LaFond[2], 2011). It has four characteristics (Fig.1.1):

1. future residents can take part in the plan and design process and they are decision-makers;

2. Co-housing projects are community oriented and internal neighborhood design is great part of the project;

3.Co-housing projects are usually equipped with extensive common facilities;

4. residents take the charge of daily management independently.

The forms and the scales of the projects are flexible, ranging from single-family houses to large community houses and from retirement houses to inter-generation houses.

The actors of Co-housing projects include those individuals, organizations or institutions that are involved in the development of Co-housing projects. Experiment-days 12 seminar[3] divided them into 3 categories: governmnet organizations, Civil Societies (non-offical and non-profit organizations) and Private groups. (Fig1.2)

1 Definition from the website: http://www.cohousing.org

2 Michael LaFond, director of id22(Institute for Creative Sustainability)

3 The EXPERIMENTDAYS is held by id22 (Institute for Creative Sustainability) annually and is a platform for cooperative housing projects, ideas and actors of creative sustainability in order to understand the city as a designed living space. Since 2003, the EXPERIMENTDAYS has offered the possibility to learn about different kinds of self-organized housing, find other supporters and to think about the future of urban development. The EXPERIMENTDAYS offered a place for ideas and actions of everyday life, art and politics to be developed through discussion, building, and experimentation. Source: www.experimentcity.net

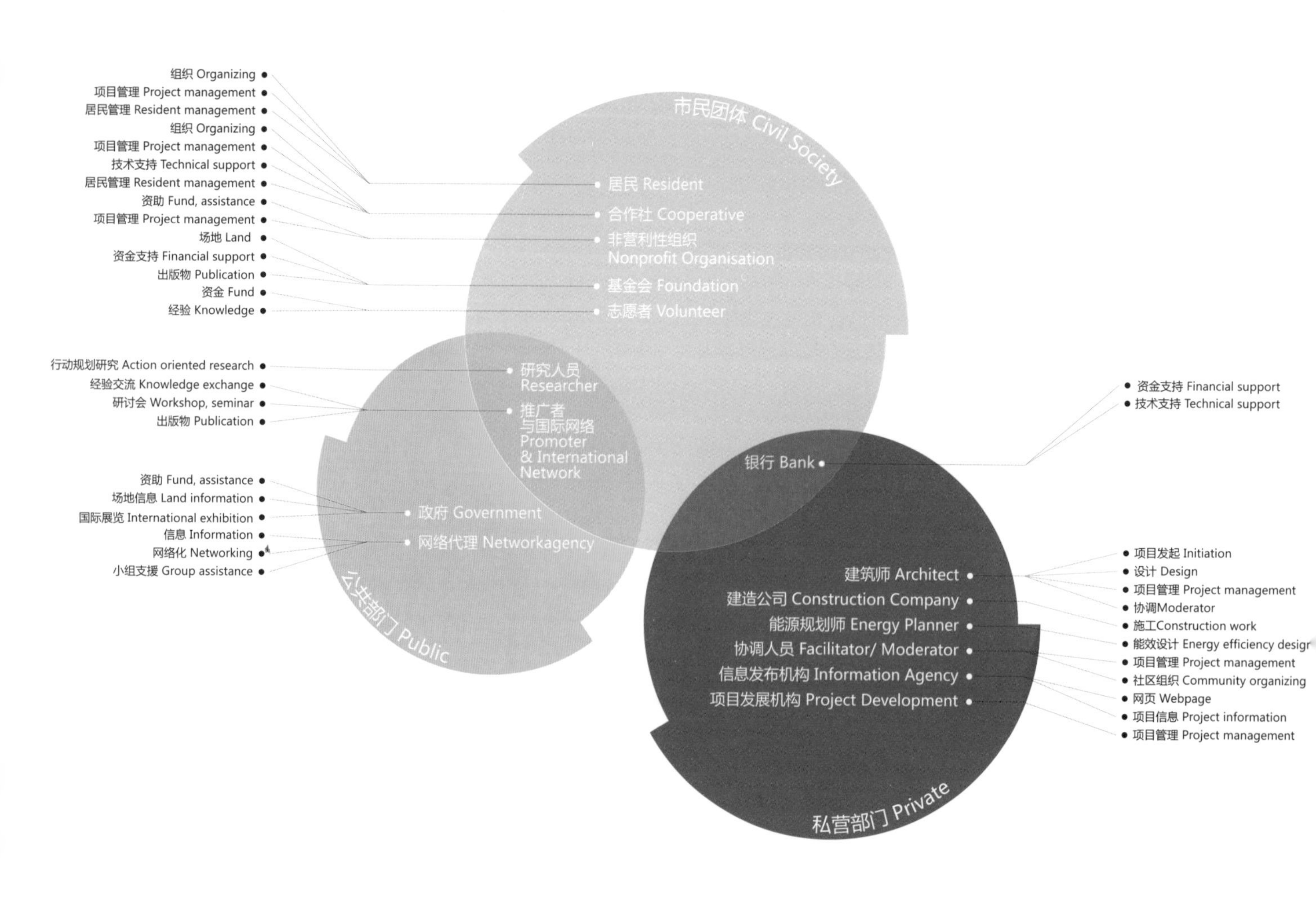

Fig.1.2 联建住宅的利益相关者 Actors of Co-housing

1.2 德国联建住宅的发展

1.2.1 联合居住理念萌芽期(19 世纪末 -20 世纪初)(Fig.1.3)

德国联建住宅的历史与德国悠久的自建文化及自我组织的传统息息相关，而这其中，“合作社”作为联建住宅项目最古老的合作形式之一，早于联建住宅项目诞生。在1840-1873 年间激烈的工业化时期里日益恶化的劳动条件下，劳动者们联合起来于 1847 年成立了第一个合作社 [1]。1862 年第一个住房合作社在汉堡成立，合作社第一次在住房领域实践。1889 年，合作社法案(Cooperative Act)正式通过。在第一次世界大战后至 1929 年经济大萧条时期，合作社成为反映劳动者基本需求而被普遍使用的社会工具和手段。

以社区生活为导向的居住理念在这一时期开始萌芽，联建住宅作为一种替代性居住模式，被认为是一种“非标准住宅”形式，而当时标准的住宅是独栋别墅和公寓，住户与住户之间联系甚少。“住房不仅意味着你头上的屋顶还意味着社区的方方面面以及公共设施”(P.ACHE,M. FEDROWITZ,2011)，20 世纪初期的住房合作社运动正是基于这一理念进行的。

这一时期基本上没有建筑方面的设计尝试，只是在传统住宅建筑中实现联合居住的理念。而随着 1930 年代纳粹废止合作社，以及受到激进的工业化与城市化进程影响，联建住宅项目蛰伏许久，直至 1970 年代才开始复苏。

1.2.2 居住形式探索实践期(20 世纪 70 年代 -90 年代)(Fig.1.3)

Novy-Huy [2] 曾指出第二次世界大战后，德国约有四百万房屋被摧毁，一千二百五十万难民流离失所。德国开始重建城市和工业，并向低收入者提供国家贴息贷款的社会

1 1847 年，F. W. Raiffeisen 创办了德国第一个救助合作社，几乎同一时间 H. S. Delitzsch 创办了德国第一个“原材料”合作社，他们都是为了保障特定行业劳动者的合法权益。

2 Rolf Novy-Huy，Stiftung Trias 德国非营利基金会负责人，许多联建项目获得该基金会扶持。

1.2 Development of Co-housing in Germany

1.2.1 Beginning (Late 19th Century-Early 20th Century) (Fig.1.3)

The history of Co-housing in Germany goes along with the development of self-made culture and self-Organization tradition in Germany. As one of the forms of Co-housing, Cooperative was born before Co-housing. Between1840-1873 the intense industrialization period, with the deteriorating working conditions, workers united together to protect their own interests and the first cooperative [4] was established in 1847 and the first home-ownership housing cooperative was founded in Hamburg in 1862. In 1889 Cooperative Act was passed and during the time from WWI to the Great Depression (1929) cooperative had become a popular social tool and instrument to reflect the basic needs of workers.

The community-oriented living concept emerged during this period and “for a long time living in a community has been an alternative to living in the ‘standard’ situation as a single family or an individual in a single house or apartment. The cooperative housing movement at the beginning of the twentieth century was already based on the perception that housing means not only having a roof over your head, but also including all the aspects of a community and its infrastructure.” (P.ACHE, M. FEDROWITZ, 2011)

During this period little architectural practice was tested and people just tried to realize the community living in the traditional housing. With cooperatives were abandoned by Nazi in 1930s and the intense process of industrialization

4 In 1847, the first aid association was founded by F. W. Raiffeisen and at the same time independently the first “raw materials association” was found by H. S. Delitzsch. They are the first cooperative in Germany and the tool to protect workers' interests.

1.2 德国联建住宅的发展 Development of Co-housing in Germany

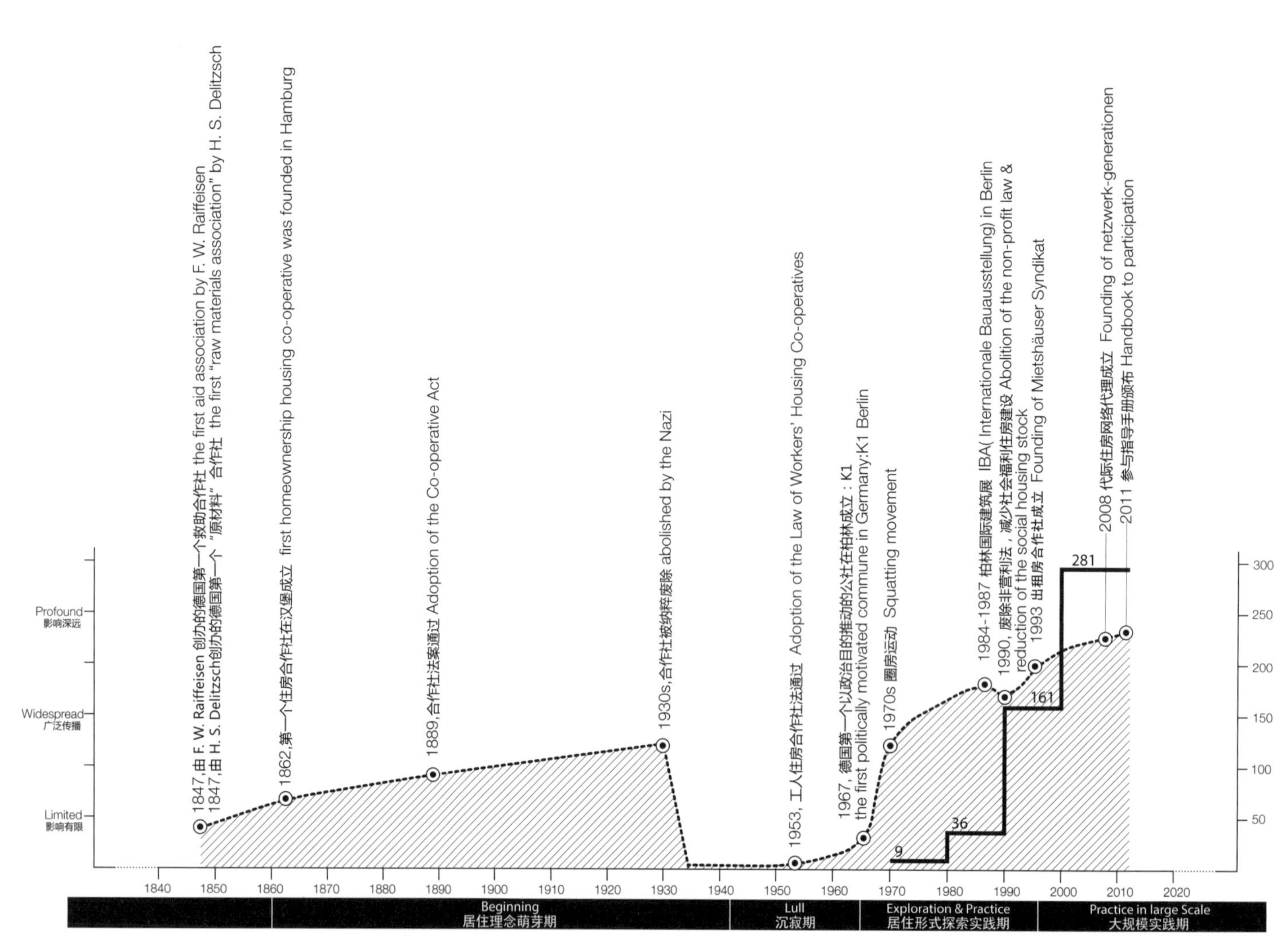

德国联建住宅相关思想与事件 Thoughts or Events related to Co-housing in Germany

1970年至2011年每十年间德国新成立联建住宅项目的数量 Number of Co-Housing's Founding in Germany in each ten years from 1970-2011

德国联建住宅相关思想与事件的影响趋势 Impact Trend of Co-Housing's related thoughts and events in Germany

Fig.1.3 德国联建住宅发展时间轴 Timeline of Co-housing in Germany

住宅。1968 年在巴黎爆发的学生运动旨在争取更大的个人自由，改革社会价值观，以及实现更多自由理念和社会生活方式。这场运动席卷了包括德国在内的许多国家，德国共同居住的生活理念以及第一个公社在这一时期形成。

1970 至 1980 年代的炒房现象引发了民众针对主管部门的抗议与暴力活动，人们纷纷占领了那些空置的、租金高昂的房屋（当时柏林约有 300 多栋住房被占领）。学术上普遍认为联建住宅也包含这一部分被占房。这一时期的城市政策倡议者尝试以集体住房和其他替代性住房模式在一起聚集生活。这些项目有的失败了，有的改变了初衷，有的则创建了植根于当地文化传统与邻里的联建住宅，这些先锋实践者为提供可负担的自管理住房实践经验做出了贡献。

这一时期各种理念的联建住宅项目开始了小范围的实践：自我管理、自我组织的联合居住模式（1970 年代）；带有共产性质的自治模式和针对妇女、丧偶老人等特定人群的联合居住模式、圈房运动（1980 年代）；社区养老模式（1990 年代）；这一时期相应的建筑主要以更新改造为主。

1984–1987 年的柏林国际建筑展（Internationale Bauausstellung Berlin，以下简称 IBA）是这一时期的经典案例。“IBA 的项目包括了城市设计、新建筑设计和旧城区的保护改造设计……，并分别提出了两个指导性原则：‘批判的重构’（kritische Rekonstruktion）和‘谨慎的城市更新’（behutsame Stadterneuerung）”（李振宇，2007，10）。IBA 的旧建筑改造部分可以说是联建住宅团体第一次真正意义上进行的建筑实践，其分布如 Fig.1.4 所示。IBA 采取了民主的、自下而上的工作方法，并于 1982 年确立了 12 条原则 [3]，归纳起来这些原则有如下几个重点：

“居民和企业主共同参加规划，更新的目标和措施应协调一

3　这 12 条原则于 1983 年 3 月在州议会上通过：（1）更新的规划和实施必须同现时的居民和企业主共同进行；（2）规划者、居民和企业主对更新的目标和措施应协调一致，技术和社会的规划必须紧密结合；（3）克劳伊茨贝克的特征必须保留，必须唤起对危旧地区的希望和信心，紧迫的房屋危险必须立即排除；（4）平面的谨慎的改变可以采用新的住宅形式；（5）新住宅和建筑应一步步出现，逐渐补充；（6）建筑状况应通过少量的拆除、街区内部绿化的补充和立面的造型来完善；（7）街道、广场、绿地等公共设施必须根据需要来更新和完善；（8）涉及的分配权和产权必须在社会规划中确定；（9）城市更新的决定必须公布，并尽可能现场讨论，涉及者的代表权必须加强；（10）获得通过的更新方案需要有明确的经济保障，经费必须尽快拨给；（11）要发展多样化的实施者，委托修复任务和建筑措施必须分开；（12）从 1984 年起，城市更新必须被保证根据这个方案进行。

and urbanization, Co-housing practices were put into a lull Not until 1970s did they thrive again.

1.2.2 Exploration and Practice Period (1970s-1990s)(Fig.1.3)

After World War II, it was estimated 4 million houses were destroyed during the war, and housing was needed by about 12.5 million refugees. Germany started the period of reconstruction of the cities and industry. The social housing with state-subsidized loans began in order to provide housing for people with low income. It made the housing affordable for the people. In 1968, the student revolt began in Paris, which demands for greater personal freedom and reform in social values with more liberal ideas and in way of life in society. This movement spread in many countries, including Germany. During this period of time, the concept of shared living arrangements and the fist commune was formed. (Novy-Huy [5], 2011, p.76-79)

The property speculation during 1970s-1980s led to the protests and violent activities against to relevant authorities. Those vacancies and high-rent buildings were occupied. Nowadays many relevant scholars think that Co-housing should also contain parts of squatter houses. Due to the inappropriate urban renewal and development policies, vacant apartment buildings, barracks and factories were occupied by the citizens and in Berlin there were about 300 squatter houses. Urban policy advocators in this period occupied the houses and tested the collective living and other alternative living models. Some of them failed, some of them changed their initial choice and others created Co-housing projects that rooted in the local culture and neighborhood, which contributed to the affordable self-organized housing development.

During this period lots of Co-housing living concepts were

5 Rolf Novy-Huy, director of Stiftung Trias

Fig.1.4 1984–1987 柏林国际建筑展旧建筑改造项目分布图

Internationale Bauausstellung Berlin 1984-1987 urban renewal projects map

致；保留特征，唤起信心；改善条件，满足需要；少量拆除，可以补入新建筑形式；城市更新的决定必须公布和讨论；给予经济保障；发展多样化的实施主体……IBA旧建筑改造项目形成了一系列规划和实施的方法（如自助式项目），对旧城改造的转变产生了极大的影响。”（李振宇，2007，16）

1.2.3 大规模实践期 (21世纪以来至今)(Fig.1.3)

在接下来的几十年里，联建住宅结合建筑标准、节能、市场导向项目、租赁房、多代居等形式创造了许多新的模式和种类[4]。联建住宅因其不同目的，取名也不尽相同：代际居、老年住宅、妇女住宅、残障人士联合住房、特殊疾病人群共享房等。回顾历史，联建住宅在德国的发展动力与运作框架也正在发生改变，这一时期联建住宅合作社形式得以复兴并且进行了大范围的实践。

随着城市中心的小型空地使用渐趋饱和，大型开敞空地逐渐成为实践的主场地，许多小型联建住宅团体联合起来共同开发项目。由于当代人的生活需求与理念发生了很多变化，在住宅创新方面也出现了许多惊喜，这一时期与之相应的建筑以新建为主。

这一时期具有划时代意义的案例是慕肯社区项目[5]（Möckernkiez e.G.），项目占据了整个城市街块，上百人规模的合作社共同决定小区的设计 (Fig.1.5)，其多维度可持续的理念整合了来自社会、环境和经济等方面的内容。

为了推广联建住宅项目，一系列的公共平台也在积极宣传联建项目的优势与具体实践方法。这其中具代表性的是欧洲实验城市（Experimentcity Europe），这是欧洲联建住宅项目的交流推广平台，该平台是由柏林可持续性创意研究所（Institute for Creative Sustainability, 简称 id22）发起并协调组织的，该平台出版了一系列的联建住宅相关手册和书籍，并且每年组织名为实验日（Experimentdays）

4 来源：Droste,C. And Siedow, T., K.（2012）Politics and cultures of community-oriented, self-organized housing in Europe - typologies, orientation and motivations, 29

5 实例7

tested in a small scale: Living in a community became a model again in the 1970s, with a strong focus on self-regulation and self-organization.More commune projects emerged in the 1980s .The 1980s was also the decade of Co-housing-projects which focused on special target groups, e.g. women or single parents. In the 1990s a new focus on projects designed for older people appeared. (P.ACHE, M. FEDROWITZ, 2011) In this period, the related architectural practice focused on transformation and renewal.

1984-87 IBA Berlin (Internationale Bauausstellung Berlin) was the typical case of this period. IBA proposed two guiding principles: “critical reconstruction” and “gentle urban renewal”.

The transformation projects of the old buildings in IBA can be clarified as the first architectural practice of Co-housing groups and its projects are located as Fig1.4 shows. IBA adopted the democratic bottom-up working method and established 12 basic principles for the “gentle urban renewal”. From then on, inhabitants readily take part in the process of their housing renewal programs and other civic issues.

1.2.3 Large Scale Practice Period (21st Century-) (Fig.1.3)

As most of the small parcel lots within the city center have been developed and the sites of new projects start to turn to big open areas. Many small Co-housing groups unite together and develop a big area together(Fig.1.5). With the change of contemporary living demands and concepts, Co-housing projects have created many surprising innovation. Many new Co-housings are built in this period.

During this period Möckernkiez e.G.[6] project has the epoch-making significance. It occupied a whole city block with

6 Case 7

Fig.1.5慕肯社区规划讨论会现场 Möckernkiez e.G. planning workshop

的公众开放活动，带领公众参观一些已经建成的联建住宅项目，并且开设论坛，旨在将联建住宅这种“建造家园”的方式带到更多人的身边。

这一时期，联建住宅项目对于城市建设、社会焦点等方面产生的积极影响受到了政府的关注，并获得了政府在经济及政策上的支持。同时，受柏林参议院城市发展部门[6]委托成立的代际住房网络代理[7]于2008年成立，旨在向市民提供联建住宅项目的咨询服务。2011年柏林参议院城市发展与环境部门[8]颁布了《参与指导手册》[9]，帮助市民参与城市发展建设的相关决策过程。这些都极大地规范和保证了包括联建住宅项目在内的自建项目的实践与运营。

6 Senatsverwaltung fü r Stadtentwicklun

7 Netzwerkagentur GenerationenWohnen

8 Senatsverwaltung fü r Stadtentwicklung und Umwelt

9 Handbuch zur Partizipation

30,000 m² and hundreds of people take part in it. They aim at creating an ecologically sustainable, accessible, cross-cultural and socially inclusive community.

In order to promote Co-housing, a series of public platforms are established to propagandize the advantages of Co-housing projects and their practice ways. Among them is “Experiment-city Europe”, which offers EuropeanCo-housing projects communication platform. It is initiated by id22（Institute for Creative Sustainability）. It emphasizes the development of a diversity of cooperative, environmental and affordable housing forms as an essential contribution to a culture of sustainable urban development. With project exhibitions (EXPERIMENTDAYS), Internet databases (WOHN-PORTAL), publications and various events such as workshops, study visits (Creative Sustainability Tours), exhibitions or Wohnsalons, experimentcity communicates the potential of a culture of creative and participatory sustainable urban development. (source: http://id22.net/en/experimentcity/)

During this period, positive influence of Co-housing on the urban development,social focuses and other aspects gradually wins attention of government and thus the support of relevant policies. Meanwhile commissioned by the Berlin Senate for the urban development [7], Network agency for the intergeneration-housing[8] was established in 2008 aiming at providing relevant consultation service on Co-housing. In 2011 Berlin senate for the urban development and environment [9] published “Handbook to participation” [10], helping citizens take part in the relevant decision-making process of urban development and other civic issues. All these efforts greatly regulate and guarantee the practice and implementation of Self-made projects including Co-housing.

7 Senatsverwaltung für Stadtentwicklung

8 Netzwerkagentur GenerationenWohnen

9 Senatsverwaltung für Stadtentwicklung und Umwelt

10 Handbuch zur Partizipation

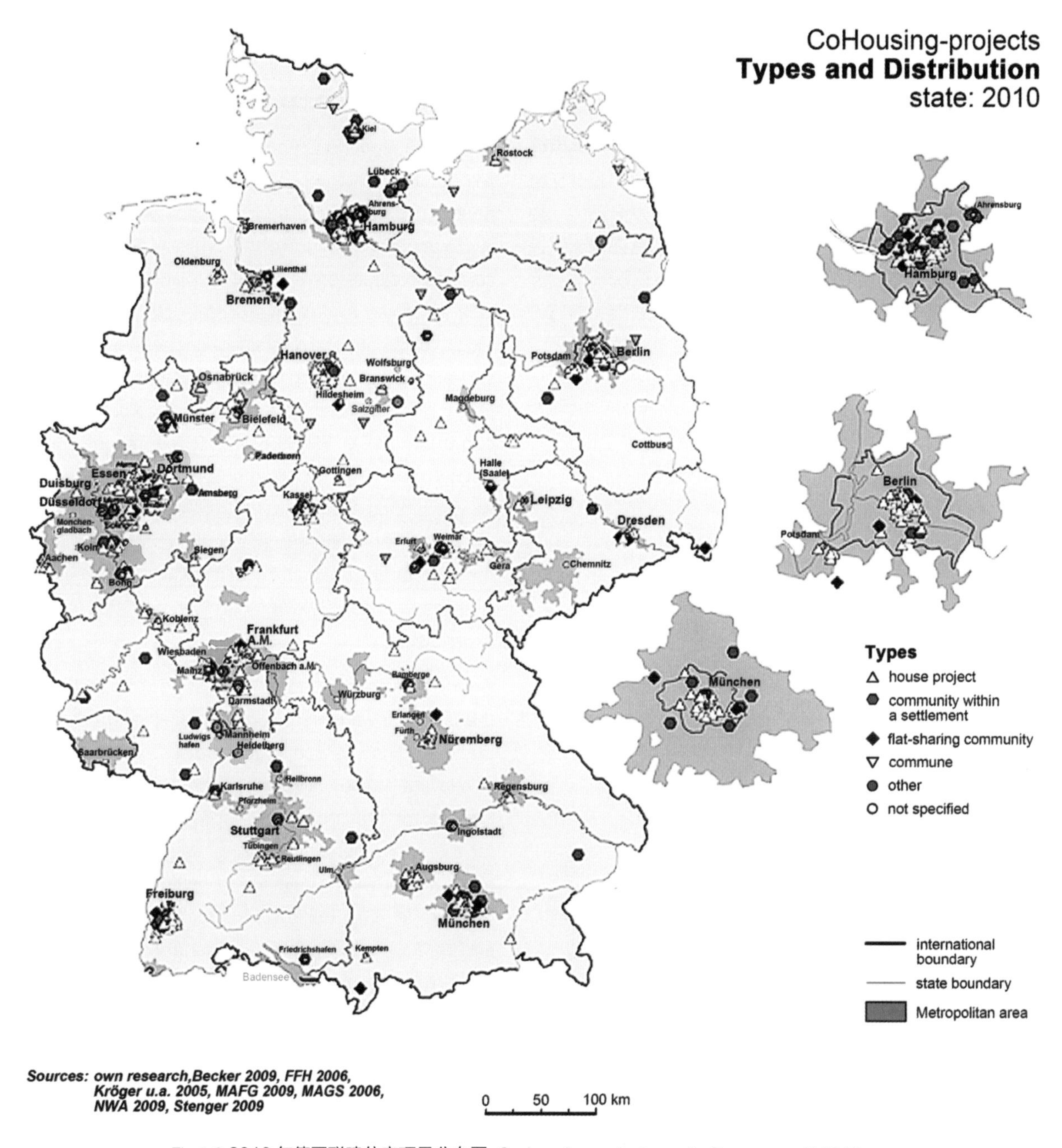

Fig.1.6 2010 年德国联建住宅项目分布图 Co-housing project map in Germany until 2010

1.3 柏林联建住宅的现状

联建住宅项目近年来在德国十分流行，由德国联建住宅研究学者 Micah Fedrowitz 绘制的分析图 (Fig.1.6) 展示了这种生活形式存在的区域以及联建住宅项目实践类型的组成。

1.3.1 联建住宅在德国的分布

截止至 2011 年 9 月，德国有超过 500 个联建住宅项目，大约居住了近 22500 人。从 Fig.1.6 中可以清楚地发现这些项目聚集在汉堡、柏林、慕尼黑和鲁尔区以及汉诺威、弗莱堡和莱茵美茵区域。这些地方曾经都是德国“圈房运动”[1]的发源地，拥有悠久的自建文化历史。Fedrowitz 指出在这些区域，居民虽然在寻求社区生活的新形式，但也不愿意离开城市生活，所以他们聚集在城市中实践可替代性的现代生活方式。据统计，联建住宅项目以小型规模组织在一起（50% 的项目成员数少于 30），70% 的项目是独栋住宅或拥有独户隔间的建筑，20% 的项目是小型社区，8% 的项目是公社或共享公寓等紧凑型建筑，公社通常位于农村而共享公寓则位于城区 (Ache and Fedrowitz, 2011)。 据 Kristian Ring 在“Selfmade City Berlin: Stadtgestaltung und Wohnprojekte in Eigeninititie”一书中统计，截止至 2012 年 10 月，有超过 125 个联建项目在柏林实现，分布如 Fig.1.7 所示。

1.3.2 联建住宅在柏林流行的原因

Fig.1.3 还显示了从 1970 年至 2010 年德国联建住宅的发展，可以看出其发展之迅速。在这一背景下，柏林也正在经历着一场前所未有的联建和集体式居住热潮，与传统的房地产开发项目相比，他们拥有明显的优势：用联建的方式建造相同或更高质量的房屋，花费要比从开发商那购买便宜 25%（开发商还面临楼盘宣传阶段的各项花销、资产所得税以及利润与风险附加税等），此外联建团体成员还能获得高度订制且生态环保的住所。在这些领域的专家诸如建筑

1 1970 至 1980 年代，德国由于日渐猖獗的炒房投机，最终导致的民间向主管部门发起的抗议与暴力活动，在这一运动过程中，人们纷纷占领了那些空置的或是无法承担租金的建筑。

1.3 Current State of Co-housing in Berlin

1.3.1 Co-housing's distribution in Germany

"Co-housing initiatives are becoming more and more attractive in Germany. An empirical study of the phenomenon (Fedrowitz, 2011) counts about 500 projects with more than 20,000 inhabitants. This is a large number compared with other countries, but compared with Germany's 40 million households and 80 million inhabitants (2010), the figures are still marginal. However, Co-housing projects seem to be on the rise..."(P.ACHE, M.FEDROWITZ, 2011) From the Fig. 1.6 we can find that major Co-housing projects assembled in Hamburg, Berlin, Munich, Hanover, Freiburg and other big cities and regions. Those areas are the origins of "Squatter Movement [1]" in Germany and citizens there have long tradition of self-made culture. They are ready to explore new ways of Community lifestyles within the city. According to statistics, "almost 70 percent of the projects are house-projects consisting of a single house or building, where each house-hold has a separate flat, and 50 percent of the projects are inhabited by fewer than thirty people. Slightly more than 20 percent of the projects are communities within a settlement, consisting of several houses with one or more flats each. The types of projects with the most intense community life, like communes and flat-sharing communities, account for only about 8 per cent. Communes are often located in rural areas, as this enables a holistic self-contained lifestyle and provides the basis for a self-sustained agriculture. Flat-sharing communities on the other hand, where all residents live in a shared apartment, are more frequent in an urban context." (Ache and Fedrowitz, 2011)

As Kristian Ring has done a comprehensive research on

1 In 1970s and 80s the property speculation resulting protest and violent toward the authorities and squatters occupied buildings, which left vacant or made unaffordable. (Novy-Huy, 2011, P.79)

1.3 柏林联建住宅的现状 Current State of Co-housing in Berlin

Fig.1.7 柏林联建住宅项目分布图 (截止至 2012 年 10 月) Co-housing in Berlin until 2012.10

师、项目经理等在项目的准备阶段就与联建团体配合工作，帮助把关，使得项目运行的流程变得更加专业和标准化。在许多专家眼中联建住宅模式是最佳的城市建设方式之一：首先，这种模式使人不必从城市搬往郊区，将具有较强经济能力的群体绑定在城市之中，从而保证城市的税收；第二，在最初的协作规划阶段，即便是拥有些许缺陷，在开发商看来需要冒风险的地块与房型也能够找到它的适合者；第三，由于联建住宅团体在最开始就考虑了他们入住后何种运营花费和因素将在日后生活中起到决定作用，因而他们尽可能地采用生态、节能的建造方式，这使得许多联建住宅项目成为生态建造方面的示范；第四，确保住房租金和产权花费的稳定性和可负担性是规划养老中最重要的影响因素。人们能够获得某地永久的居住权将有效地减少政府对于公众退休福利的外部投入，因此联建住宅项目不仅能够以节能方式建造房屋以降低各项辅助成本，同时也能为养老提供积极准备。出于以上原由，柏林政府越来越关注该种实践方式，从政策方面大力扶持，为联建住宅项目的发展提供了有力的外部环境。

1.3.3 柏林联建住宅的实现方式

柏林联建住宅项目有三种所有制模式及五种实现形式。所有制模式分为私有模式，合作社模式以及租赁模式(Fig.1.8)。五种实现形式分为合作社、有限公司、协会、民法组织及其他，各种形式对于人员规模、资金以及组建时间各有不同要求，基本上涵盖了可能出现的情况（Fig.1.9）。

the existing Co-housing projects in Berlin, we can find that over 125 Co-housing projects have been initiated in Berlin by October 2012 and this number is still on the rise. They distribute largely within the S-Bahn ring as Fig.1.7 shows:

1.3.2 Reason for the popularity of Co-housing in Berlin

Fig.1.3 also shows the numbers of newly founded Co-housing projects in Germany every ten years from 1970 to 2010. We can find that the numbers of Co-housing are on the rise recently. Under this context, Berlin is also undergoing an unheard of surge of Co-housing and collective living heat. Compared with the traditional real estate development projects, Co-housing has the obvious advantages. By Co-housing approaches, people can get houses of the same even better quality than those from the traditional housing market with 25% less money. Because usually the developers have to pay more on the promotion and a series of taxes. What's more, Co-housing group can achieve a highly customized and Eco-optimized living place. That's the major point why Co-housing projects continually attract lots of people. Additionally the process of projects nowadays is becoming more and more standardized and professional, with the help of the relevant experts, the new-comers are more easily to initiate a project.

In the view of many experts, Co-housing is no doubt one of the best urban construction approaches. First, "this model does not entail parties moving away from the city into the countryside, keeping tax revenue within the city and more closely tying a financially strong clientele into the city" (Härtel, 2007); Second, "during the phase of collaborative preliminary planning, even properties with a difficult layout that real estate developers would slot as risky find their takers" (Härtel, 2007);Third, "since communal living and Co-housing groups consider from the very first moment what

1.3 柏林联建住宅的现状 | Current State of Co-housing in Berlin

私有模式 Private Property Model

在这种模式下，居民即为住宅所有者，公共部分为集体所共有。相比合作社模式，集体可以以较少的资本获得贷款。

In this model, the residents are the owners of their appartments while the communal area are owned by the collective. Compared with Cooperative Model, it can get loan with less capital

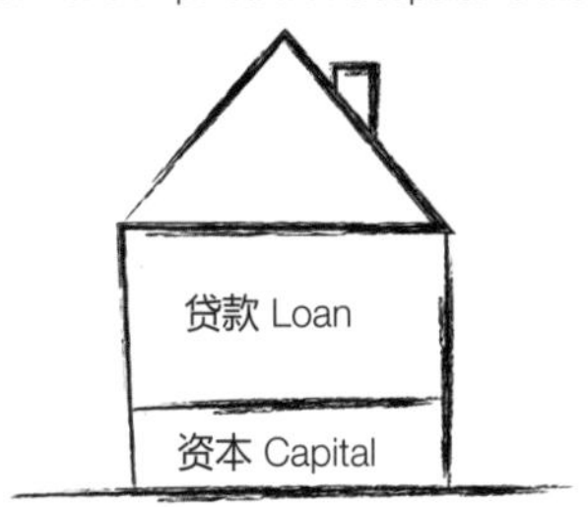

·无房租 ·No Rent
·政府支持 ·Government Support
·税收减免 ·Tax Advantage
·高品质生活 ·Living Quality
·安全投资 ·Safe Investment
·养老投资 ·Pension Investment

合作社模式 Cooperative Model

在这种模式下，住宅及宅基地归合作社所有，居民为合作社成员并享有股份。合作社模式又分为两种：既有合作社与新兴合作社。前者居民分享较多资本而非贷款压力，逐渐赢利，后者居民则承担较多贷款压力。

In this model, the property belongs to the cooperative while the members have the shares. This model can be subdivided into two: Existing Cooperative Model and Young Cooperative Model. The former share more capital instead of loan while the latter bare more loan pressure.

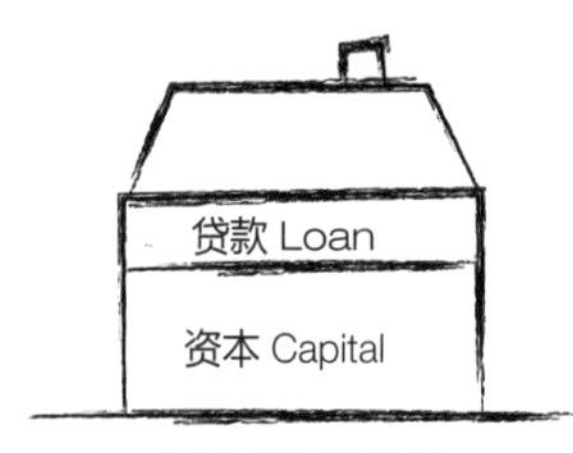

既有合作社模式
Existing Cooperative Model

新兴合作社模式
Young Cooperative Model

·融资 ·Financing
·政府支持 ·Government Support
·灵活 ·Flexibility
·较少开支 ·Less Expenditure
·安全投资 ·Safe Investment
·无债务 ·No Loan

租赁模式 Rental Model

在这种模式下，居民只是租住，住宅归他人所有。

In this model, the residents are just tenants.

·灵活 ·Flexibility
·无风险 ·No Risk
·无资本约束 ·No Capital-restriction
·不需要缩减开支 ·No Consuming-curtailment
·较少开支 ·Less Expenditure
·无债务 ·No Loan

Fig.1.8 柏林联建住宅三种所有制模式 Co-housing's Ownership

running costs and various other factors will come into play once they live in the space, they drive forward ecological and resource-efficient methods of construction" (Härtel, 2007). This leads many projects to be pilot demonstration of Eco-buildings;Fourth, "it is a well-known fact that ensuring the stability and affordability of home rental or ownership costs is one of the most important contributing factors in planning for retirement. Cooperatives and home ownership projects and all other organizational forms that enable people to acquire the permanent right to live somewhere – can reduce the need to draw on external public retirement benefits. Hence, building cooperative housing in a resource-efficient way, so as to lower ancillary costs, is also a proactive provision for retirement" (Härtel, 2007) Out of these causes, Co-housing draws more and more attention from the Senate and gains its political and economical support which creates positive external condition for the development of Co-housing in Berlin.

1.3.3 Approaches to Co-housing in Berlin

As to the ownership of Co-housing, there are three models: Private Property Model, Cooperative Model and Rental Model. Their features have been shown in Fig.1.8. There are major five legal forms for people to form Co-housing Group: Cooperative, Limited Liability Company, Association, Civic Law Partnership and others. Each form has different requirements to the least number of members, capitals and processing time almost covering all the possible conditions(Fig.1.9) .

合作社(Genossenschaft) |Cooperative

是由有着共同的经济和社会利益的人员组成的联合体。合作社的每个成员不用考虑其所占有的股份多少而拥有相同的独立的投票权，成立合作社需要2-3个月，最少需要三人才能成立，无需启动资金。合作社是以集体的名义占有房屋所有权，成员只拥有房屋使用权。
The Genossenschaft or cooperative is a union of members with the same economic and social interest. Each member of the cooperative has the same right to vote independently without consider how much share he or she has. About two until three months are needed to set up the cooperative, capital money is not needed and at least three people as a member of the cooperative.

有限公司(GmbH) |Limited Liability Company

强调成员并不以个人形式承担公司债务。有限公司的成立需要2个月，并需要缴纳25000欧元，一人即可成立。
The GmbH or a company with limited liability is a legal entity in Germany, which emphasizes the fact that the owners (Gesellschafter, also known as members) of the entity are not personally liable for the company's debt. About two months are needed to set up the GmbH, about 25,000 Euro is needed for capital and only need one person in the structure to set up the GmbH.

协会(Verein) |Association

是由一群拥有相同目的非盈利背景的人员组成的志愿组织。协会成立需要两个月，无需启动资金，至少需要七名成员。
Verein is a voluntary union of the member of people who have a common purpose with non-profit background. At least, a few months are needed to set up the Verein; capital money is not needed and at least seven people as a member of the Verein.

民法组织(GbR) |Civil Law Partnership

强调每个成员都有义务承担公司债务。民法组织的成立需要2个月，无需启动资金，至少需要两名成员。
GbR is the basic form of a company which owned limited legal ability. Every stake holder is responsible for the debt of the company. About two months are needed to set up the GbR; capital money is not needed and at least two people in as a member of the GbR.

其他：业主委员会(WEG) | Homeowner Association
有限合伙（KG）|Limited Partnership

业主委员会的成员独立拥有相应房屋所有权并同时共享公共部分所有权。
Members of WEG enjoy the ownership of their rooms and share the ownership of the commom area.

Fig.1.9 柏林联建住宅项目的实现形式 Co-housing's Approach

Typologies of Co-housing in Berlin Part 2

柏林联建住宅的类型

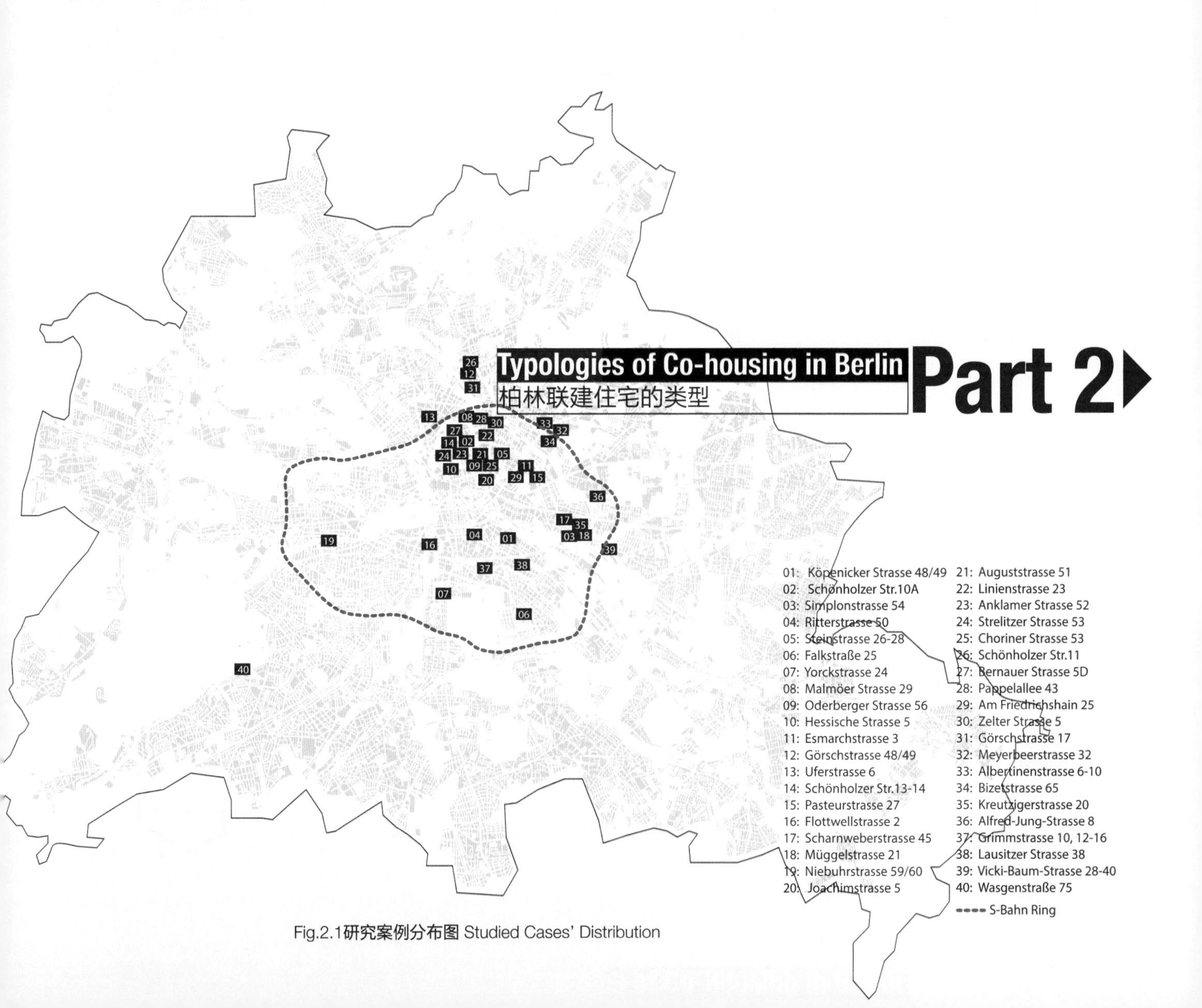

Fig.2.1研究案例分布图 Studied Cases' Distribution

Client
委托方

Architect
建筑师自我委托

Architects
Residents

Co-housing Projects

Future Resident
居民委托

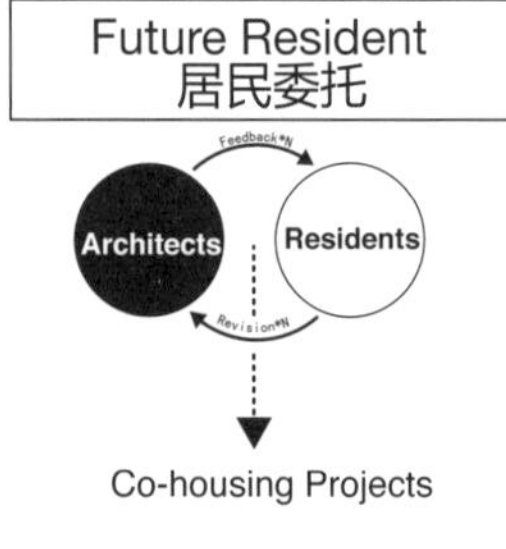

Co-housing Projects

Government
政府委托

Architects
Govern-
ment

Co-housing Projects

Public/No-Profit Entity
非营利性团体组织委托

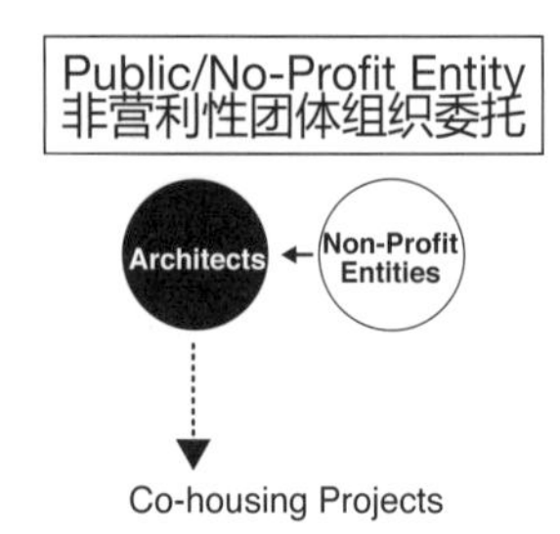

Co-housing Projects

Site Type
场地类型

Embedded Field
嵌入式小型地块

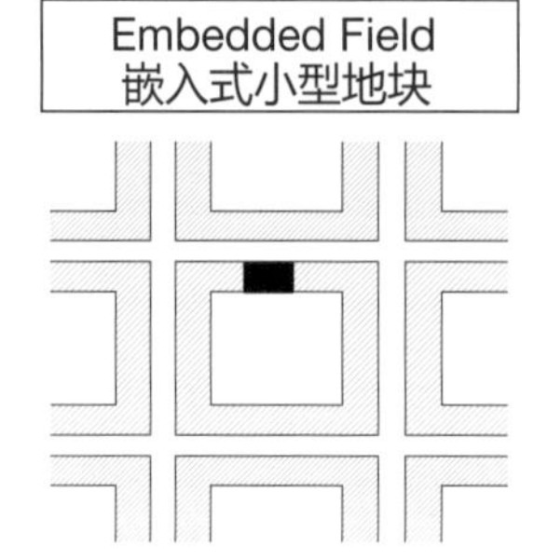

Open Field
开敞式大型地块

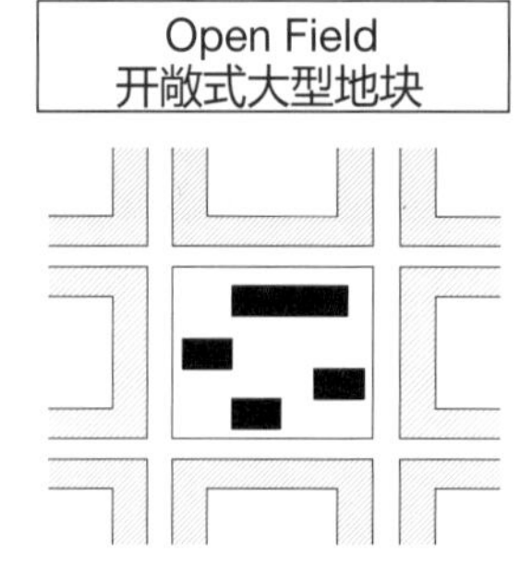

Renovated Field
基于既有建筑的修
复改扩建式地块

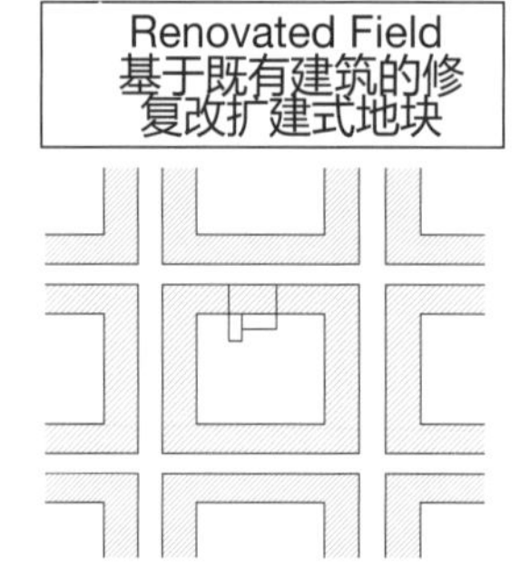

Space Organisation
空间组织模式

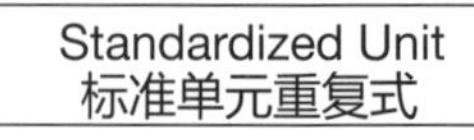

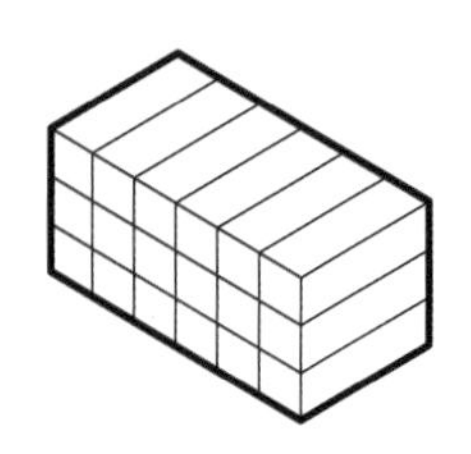

Non-standardized Unit
非标准单元组合式

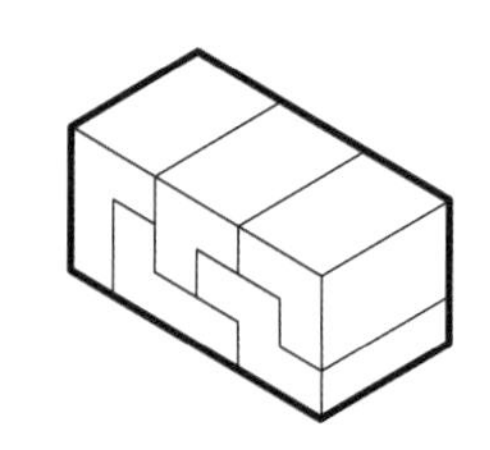

Cluster
集群式

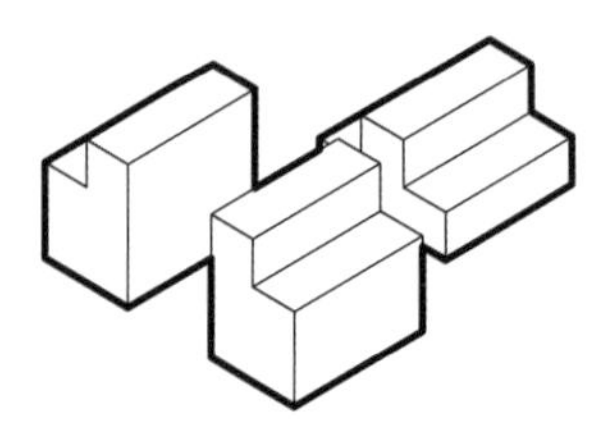

Fig.2.2 **联建住宅的分类** Co-housing's Typologies

联建住宅在其发展过程中形成了多种类型，特别是在柏林这样一座先锋城市里，由于 1970 年代反对房地产投机的“圈房运动”，城市中心区域存有大量空地、无人区和年久失修的房屋，它们无形中为项目实践者提供了丰富的空间类型与资源。人们去占领、改造和更新，在空间、建筑、起居和工作等方面创造了柏林建筑的多样性与高品质。本章将基于 2000 年之后的联建住宅项目从社会、城市设计、建筑设计三个方面梳理柏林联建住宅的类型 (Fig.2.2)。

2.1 基于项目委托方的分类

作为项目的委托方，他们对于将来居住生活有各种畅想，但是如何能将其落到实地，还需依靠建筑师的协助。据此，联建住宅可分为四类：建筑师自我委托型、居民委托型、政府委托型和非营利性团体或组织委托型。

2.1.1 建筑师自我委托型

在这种类型中，建筑师既是项目设计者，又是项目使用者，免去了设计师与业主之间理念的沟通，能够最大限度地实现居住者的想法。联建住宅项目有很多都是建筑师直接发起、设计并实施的。这种类型存在两种情况：一种是建筑师即未来入住居民，另一种情况是建筑师根据自己的想法先行设计住宅，完成后带着方案再去寻找适合该方案的居民加盟。这类项目的特点是设计效率高，项目实施较顺畅（1~2 年），业主满意度高。

2.1.2 居民委托型

在这种类型中，未来居民掌握了发言权，建筑师负责将居民的生活理想转译成建筑实体。这类项目的特点是设计效率较低，设计方案容易反复讨论，设计阶段耗时较长，项目实施较缓慢（2~3 年甚至更长时间），但往往业主满意度很高。

2.1.3 政府委托型

柏林联建住宅项目脱胎于柏林自发项目，经过这些年的发展，它们逐渐得到政府部门的认同并获得政策上的扶持。

In its development period Co-housing has formed various typologies. Especially in such a pioneer city as Berlin there are many vacant, no-man's lands; old buildings are out of repair in the city center mainly due to the “squatter movement” in last centuries. They have provided the practitioners with plenty of potential space resources waiting to be occupied,transformed and renovated. With their participation, buildings in Berlin have achieved diversity and high quality. In this chapter the author will clarify typologies of Co-housing in Berlin from social,urban,architectural aspects respectively. In these three aspects the author will base on the selected cases to provide a horizontal understanding of current situation in Berlin(Fig.2.2).

2.1 Classified according to Clients

As the Clients of Co-housing projects, they have many wonderful ideas of future living and in order to turn them into realities, they need the help of architects. According to whom the architects are commissioned by, Co-housing projects can be classified into 4 categories: Architect Self-commission type, Resident Commission type, Government Commission type and Non-profit Entity Commission type.

2.1.1 Architect Self-commission

In this type , architects play both the roles of Party A and Party B. They are not only the project-designers but also the project-users, thus this avoiding the idea-communication between designers and clients, which may lead to translation error. Many Co-housing projects are initiated directly by architects. They are in charge of design and implementation. There are two situations in this type: one is the architects are the future residents and the other is the architects firstly design the house according to their ideas and then find appropriate people to form the groups to live in. The Characteristics of this type is that the design-efficiency are

这类项目通常主题突出，目标人群明确，设计效率高，实施较快（1~2 年）。

2.1.4 非营利性团体或组织委托型

当独立的个体一时难以寻找到合适的建设伙伴或是额定数目的合伙人时，他们往往求助于专业的社会团体或组织，加入他们认可的既有联建住宅项目。这类团体组织中有代表性的是出租房合作社（Mietshäuser Syndikat [1]）。这是德国一个独立的非营利性组织控股公司，成立于 1993 年，它旨在帮助、投资和发起自组织项目（包括联建住宅项目），为项目成员提供长期稳定的廉价居住空间。截止至 2014 年 4 月，该组织在德国拥有 82 个住房项目，其中柏林有 12 个项目（M29[2] 项目 即是该组织的成员）[3]。

这类项目运作机制较成熟完善，设计效率高，项目实施较快（几个月至一年），一般以出租为主。

2.2 基于项目场地的分类

根据场地类型，联建住宅分为三类，前两类是在空地上进行的项目实践，即嵌入式小型地块和开敞式大型地块，第三类为基于既有建筑的项目实践。

2.2.1 嵌入式小型地块

柏林参议院于 2002 年起削减了住宅项目的资金，随着这一决策的实施，德国经济一片萧条，投资商立马停止了对于住宅的投资，这直接造成很多项目烂尾，导致在城市中心出现了许多小型的建筑空地。许多人，尤其是想要住在内城的家庭，在以租房为主导的市场上找不到他们想要的居住房型，于是对于投资建设自己的公寓产生兴趣。建筑师们发现了这一潜在市场，开始为空地设计住宅项目，然后寻找足够多的合作者将其买下并建设。

这类项目的特点是继承了传统街区建筑的划分尺度，适于建设十人左右的小型联建团体的住宅项目。这类建筑通

1 网址：http://www.syndikat.org/en/
2 实例 8
3 资料来源于 Mietshäuser Syndikat 官方网站：http://www.syndikat.org/en/

relatively high; projects progress is smooth and the residents are satisfied with the result.

2.1.2 Resident Commission

In Co-housing projects, the residents have a say and they are the Party A. Architects help them to translate their ideas into architecture entities. The Characteristics of this type are that the design-efficiency is relatively low; design-scheme usually need to be discussed again and again with the residents; The time spent on the design is relatively long and the project progress is relatively slowly but the residents are highly satisfied with the result.

2.1.3 Government Commission

After decades' development, Co-housing projects came from the opposite side of the government and gradually received the support from the government.

Usually this type has clear concepts and target groups. The design-efficiency is relatively high and projects progress is fast.

2.1.4 Non-profit Entity Commission

When the individuals cannot find suitable partners or required numbers of partners, usually they can turn to some professional non-profit social entities for help. Among them is apartment-house syndicate [1] . This is an independent Non-profit Entity holding company, which was founded in 1993. It provides advice to self-organized house projects interested in the Syndikat's model and invests in projects so that they can be taken off the real estate market. Meanwhile it helps with its know-how in the area of project financing and helps initiate new projects. Until April 2014, It has 82 member-projects in Germany, of which 12 projects are in

1 Mietshäuser Syndikat Website：http://www.syndikat.org/en/

常只有两个立面，并需要考虑与相邻建筑的关系。

2.2.2 开敞式大型地块

柏林在 2000 年之后的几年里所形成的联建住宅团体大部分是在嵌入式场地上实践的项目，而近年来，这种场地越来越少，但市区内仍有大片的开敞空地，因而许多项目转而投向这类场地进行建设，项目规模发展越来越大。通常联建项目团体在进行大型开敞式场地开发时更像是在进行传统的房地产开发，但是掌握了小型地块开发的经验后，大型开敞式场地不再是由某个单独的项目团体独自开发，而是由多个联建团体相互协作共同完成。

这类项目设计局限性小，可以进行包括场地在内的建筑设计而不必受相邻建筑的干扰，共享的室外空间更加丰富，多以周边围合式或是集群式的形式组织建筑。

2.2.3 基于既有建筑的修复改扩建式地块

联建住宅项目不全是新建项目也有许多修复式项目，通过这种方式实现资源的再利用和可持续发展。这类项目依托于既有建筑，可能衍化出除了居住之外的其他功能。

2.3 基于空间组织模式的分类

每个联建住宅项目都是独一无二的，但其空间组织模式还是有一定的规律可循，据此，联建住宅可以分为三类：

2.3.1 标准单元重复式

在联建住宅个性化设计中一般有两种策略，第一种是一劳永逸型，即先将所有情况考虑在内，减少二次设计。具体说来，设计先从整体再到局部，综合各方面的意愿，先行设计整体构架与居住单元的大致分隔，然后将标准单元按照一定的空间语言置入，其空间组织模式即为标准单元重复式。完成前期的设计工作后，在项目运营阶段，使用人群发生变更时，也能灵活地进行房间布局的改变。

虽然单元是重复的，但通过内部分隔同样能实现不一样的空间形式。如针对单元设计一系列模块化的改造构件，

Berlin (M29 [2] has the membership).

The operation mechanism of this type of projects is sophisticated. The design-efficiency is relatively high and the projects progress is fast. They are usually for rental.

2.2 Classified according to Site Type

Project Site Type plays an important role in space layout of the architecture and the architectural forms. According to the site type, Co-housing can be classified into 3 categories. The former two categories aims at the buildings built on the vacant space: Embedded Field(small) and Open Field(large). The last category aim at projects based on existing buildings including transformation, restoration and extension. Here summarized as Renovated Field.

2.2.1 Embedded Field

In 2002 Berlin Senate started to reduce the capital input of housing projects. With the implementation of this policy, many developers stopped the investment in housing, which directly led to many unfinished buildings and blank areas in the city centers. The size of these space is relatively small and usually the embedded field in the existing city block, which is suitable for small size of Co-housing groups.

Many people especially those families who want to live in the inner city cannot find the ideal housing type from the rental-housing-dominant market. More and more people are interested in investing to construct their own apartments. Some architects found this potential market and started to find sufficient co-workers to buy the small embedded field and design it and after its completion even live there. Co-housing projects provide innovative solution to particular situations while the group members living there enjoy the benefit from the community lives.

---

2 Case 8

住户可以根据喜好，通过相互搭配而产生不同组合。一般这种情况下，户型的灵活度较高（自由的平面），能够根据特定需要进行再次分隔改变。这种策略使得建筑能够被反复利用，在建筑使用方进行更替时适应性强。

2.3.2 非标准单元组合式

联建住宅个性化设计的另一种策略即各个击破型。设计先从局部再到整体，建筑设计师针对每户的需求进行设计，然后综合考虑所有户型进行整合。这种模式设计成本较大，但高度贴合住户需求，不过当建筑使用方更替时，其适应性不及第一种策略。

该种类型主要出现在集约型建筑中，建筑一般受限于地块大小，主要呈垂直发展。为了提高空间利用率，不同建筑空间如积木一般相互咬合，空间类型丰富，功能较为复合。

2.3.3 集群式

近几年来，联建住宅项目不再以单栋为主，而是采用集群模式。建筑内部空间的组织仍旧遵循上述两种模式，建筑外部空间的组织即是集群式的又一关注点。

该种类型是前述两种类型的复合体，集合了所有类型特征，一般出现在大型联建住宅项目之中，功能往往比较复合。在外部空间的组织上一般有三种：

（1）**周边围合式**：遵循传统街块形式，建筑紧贴建筑红线，在内部围合形成较大型的中央庭院空间。

（2）**散落式**：建筑呈散点式布局，街道空间与社区空间相互渗透，外部空间的共享程度较大，社区氛围较为活跃。

（3）**中央环绕式**：建筑居于街块中心，外部空间环绕四周，通过绿化、城市家具的设置，限定外部空间的共享等级。

2.4 研究案例分类汇总

本书研究的 40 个案例基于这三方面标准进行分类，结果如 Fig.2.3 所示。

This type of projects inherits the division size of traditional urban block, which is suitable for the small Co-housing groups with 10-20 people. Usually they are inserted between two existing buildings and only have two facades. Design of the buildings should take the surrounding buildings into consideration.

2.2.2 Open Field

Recent years the resource of this embedded fields are almost exhausted meanwhile there are still large open field in the city center waiting to be developed. Under this context, more and more projects turn to practice on these large open fields and the sizes of the projects become larger and larger. Usually the Co-housing development on the large open field is more like the traditional real estate development, but after mastering the developing know-how of embedded fields, Co-housing groups don't develop the large field alone but unite several groups to co-develop it.

Usually design of this type has less limitations. Without the restriction of neighbor buildings, the whole site can be well designed with abundant outdoor shared space. Usually there are more than one buildings on the site. They can choose wrapping type (traditional block way) or cluster type to organize the buildings.

2.2.3 Renovated Field

Type of these Co-housing projects is based on the existing buildings. The design focuses mainly on the optimization of energy-consumption or the consideration of historic context. Usually it will develop many functions other than living.

2.3 Classified according to Space Organization

Every Co-housing project is unique, but we can still summarize some rules of the space organization models from the unique buildings. According to the space Organization, Co-housing can be classified into 3 categories: Standardized Unit type, Non-standardized Unit type and Cluster type.

2.3.1 Standardized Unit

In the individualized design of Co-housing there are usually two strategies. The first strategy is Design from whole to parts: firstly the designer integrates wishes of all parties and design the frame of the building and the general division of living units. Then put the standardized units into the frame according to the space rules set by the architects. Buildings designed through this way can be classified as Standardized Unit Type. When the users change in the future, the layout of the room can be flexibly changed without many efforts.

Although the units are repetitive but they can realize different space forms by different inner partition ways. The method can be described as follow: the designer will design a series of modular components for the standardized unit. The residents can choose them freely according to their willing to subdivide the inner space. With this strategy, the buildings can be recycled well and have strong adaptability.

2.3.2 Non-standardized Unit

The other strategy of Co-housing individualized design is Design from parts to whole:

Firstly the designers design the individual unit according to each user, and integrate the non-standardized units together to form the whole building. Usually the design cost and construction cost of this strategy is higher than the former one, but it is highly fit the users. When the users change in the future, it has much less adaptability than the former one. Building built out of this strategy can be classified as Non-standardized Unit Type.

This type is often used in the space-intensive buildings. The architectures are confined to as small site and seek to space vertically. In order to improve the space utilization and balance the contradiction among space demand, space quality and space occupancy, different space units are designed as "toy bricks" to interlace with each othert so as to create abundant space typologies and multiple uses.

2.3.3 Cluster

More and more Co-housing groups choose to unite together to develop the large open field. Due to the sufficient space, projects no longer consist of one building but building-cluster. Space within the building is organized based on the strategies talked above and the outdoor space Organization is the focus of this type.

This type is the combination of the former two types and is used in the large Co-housing projects with multiple functions . The outdoor space organization ways can be summarized as three types:

(1) **Wrapping Type**: Buildings stand on the property line obeying the traditional block form and it will have a big central courtyard.

(2) **Dotting Type**: Buildings scatter in the site. Public space permeates from the street to the inner community. It will have shared outdoor space and active community atmosphere.

(3) **Center Type**: Buildings are in the center of the block and are surrounded by the outdoor space. By greenery,urban furniture outdoor space is divided into different levels of sharability.

2.4 Classification of the Studied Cases

According to the above three criteria, 40 studied cases are classified as Fig2.3 shows.

2.4 研究案例分类汇总 Classification of Studied Cases

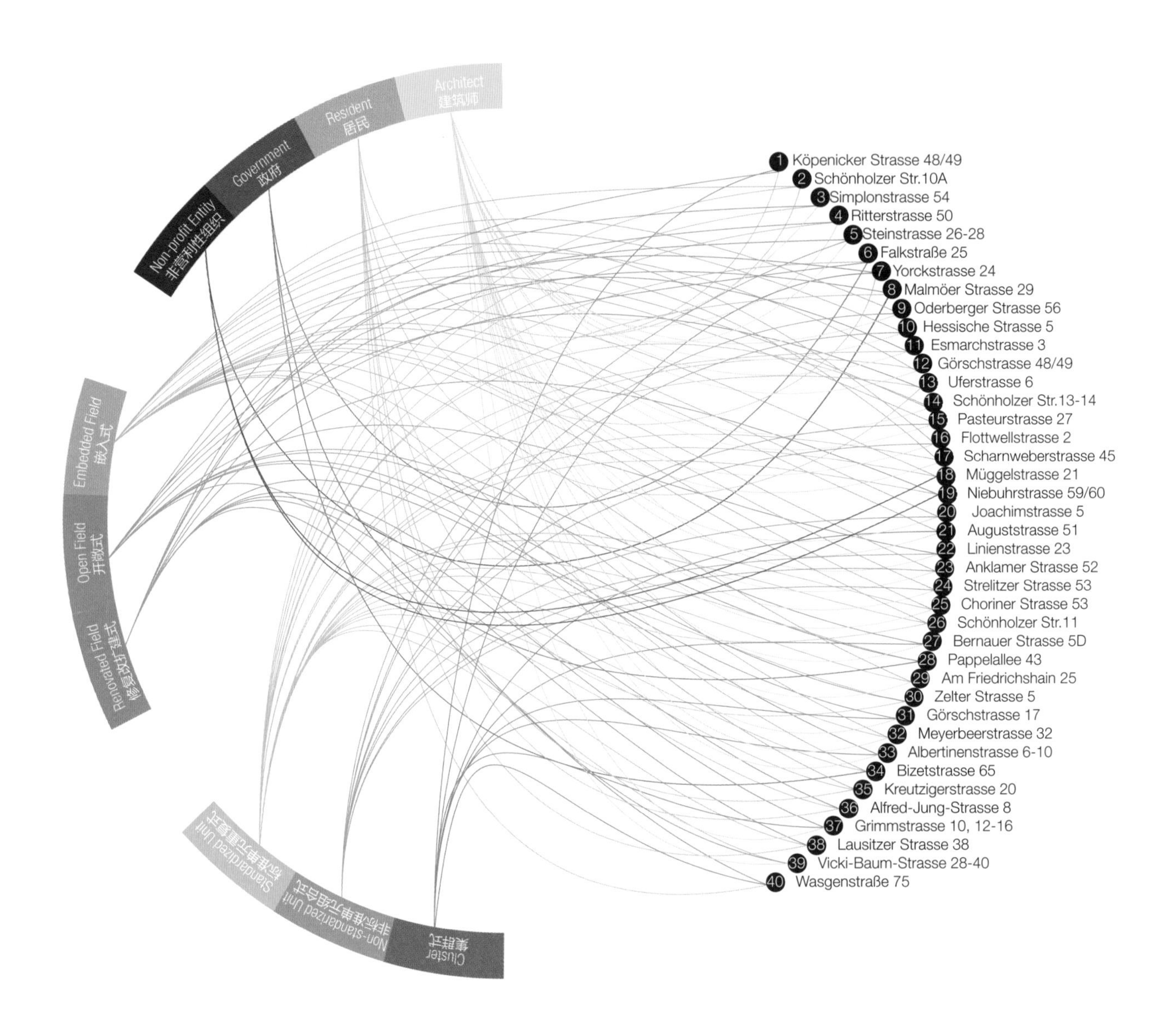

Fig.2.3 本书研究所依托的 40 个案例按照委托方、场地类型、空间组织模式分类情况汇总（前 19 个案例在本书中将做深入阐述）
40 Studied Cases in this book are classified according to Clients, Site types & Space Organization Models (the former 19 cases will be discussed in detail)

Design Features of Co-housing in Berlin Part 3
柏林联建住宅的设计特征

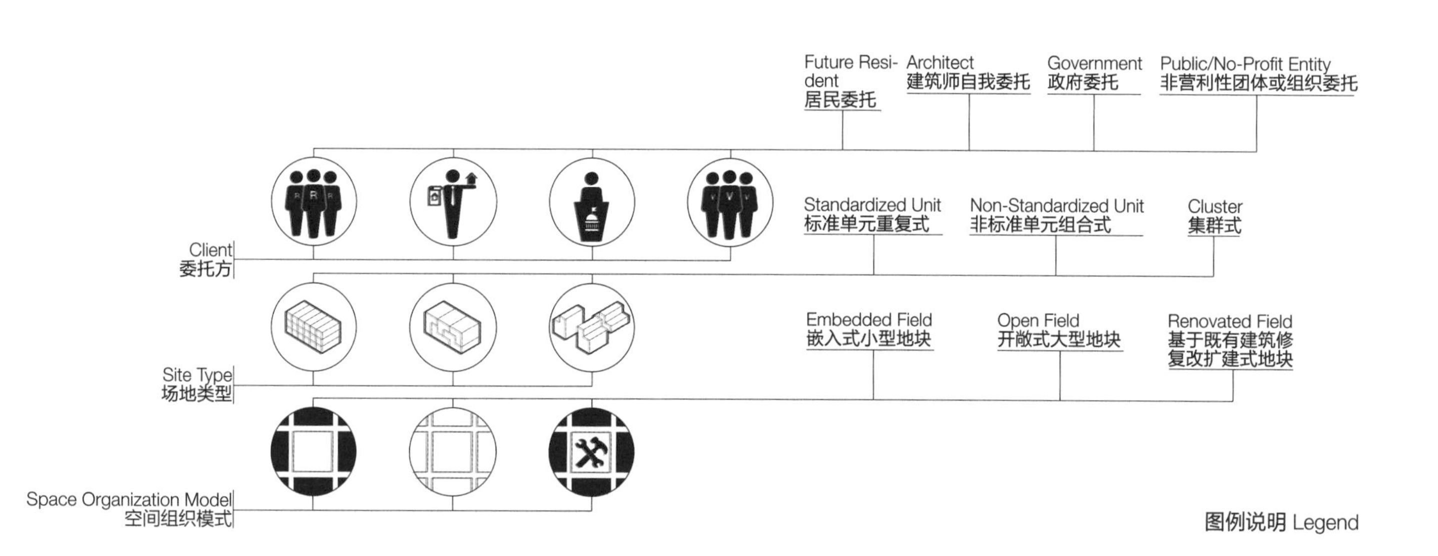

图例说明 Legend

传统住宅
室内外空间分隔明显
Traditional Housing
Clear Boundary between Indoor & Outdoor

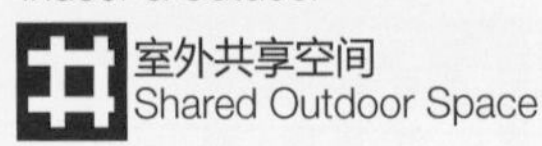

联建住宅
室内外分隔柔化
Co-Housing
Soft Boundary between Indoor & Outdoor

传统住宅
几无室内共享空间
Traditional Housing
Hardly Indoor shared space can be found

联建住宅
室内分布共享空间
Co-Housing
Indoor shared space can be found

Participation
参与性

need to participate to decide common things together and enjoy community life.

参与共同事宜决策，
共享社区生活

Sharability
共享性

Traditional Housing

Clear Boundary between Indoor & Outdoor /
Hardly Indoor shared space can be found

Co-Housing

Soft Boundary between Indoor & Outdoor /
Indoor shared space can be found

传统住宅：室内外空间分隔明显，几无室内共享空间；
联建住宅：室内外分隔柔化，室内分布共享空间。

Fig.3.1参与性与共享性 Participation & Sharability

3.1 参与 · 共享

联建住宅居民在规划设计过程中的参与性体现在参与形式的多样性、参与阶段的自主性和参与规模的灵活性这三方面。居民可以选择多种形式结成联建项目团体，自主决定参与项目进行的阶段。参与项目团体的规模从两人到百人浮动很大，项目本身的规模根据目前的实践来看，从几户的公寓式住宅到几百户的大型居住社区，其跨度也十分大。通过与规划师、建筑师、项目经理等各运作过程中的专家一起讨论，成员能够量身定做居住空间类型与生活方式，而由此产生的最具特色的空间类型为共享空间。联建住宅项目的成员渴望更多的人际关怀和生动活泼的社区生活，所以共享空间在这一类项目中占据着十分重要的地位，它们是集体生活的物化象征，是社区活动的空间基础 (Fig.3.1)。

3.1.1 外向的参与：共享的室外空间

大部分联建住宅项目团体通过长时期的讨论与合作，不仅在各成员之间，而且在团体与项目所在地之间也产生了深厚的认同感，而这种认同感进一步演化成为了一种区域使命感与社会责任感，这使得联建项目中有许多回馈社会、回馈城市、鼓励外向性参与的室外共享空间。伴随着项目本身的发展，这些共享空间也正一点一点地改变着城市的面貌。

3.1 Participation & Shareability

Co-housing group can take part in the whole process of the project. This kind of participation includes the variety of participation forms, self-decision-making on the phase of participation and the flexibility of the participation scale. There are lots of legal forms that people can choose to form Co-housing group and can freely take part in the whole or parts of the phases. As to the scale, there are two folds: one is the group size can range from two to hundreds; the other is the construction scale can range from a townhouse to a big living community. With the help of planers, architects, project managers and other professional experts, group members can discuss together and order their own ideal living space and living style. Most Co-housing members go for a more interpersonal caring and lively community life, so shared communal space occupies a very important position in the Co-housing projects(Fig.3.1).

3.1.1 Outward Participation: Shared Outdoor Space

Most Co-housing projects are spontaneous and after longtime co-work, not only among the members but also between the group and the project site a deep sense of community is slowly forming, which reflects on the shared outdoor-space. Different from the private courtyard in the normal housing, most of these spaces are open to the public and even some of them are used for urban pioneer practice for the neighborhood and for the city.

01\ Spreefeld Berlin

立体室外共享空间整合实验功能
3D shared outdoor space system with experimental function

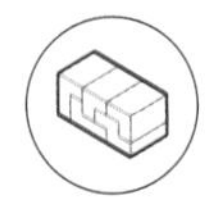
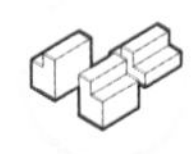

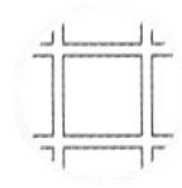

项目概况 Project Profile

项目名称 Project Name：	Spree Berlin
项目地址 Project Address：	Köpenicker Strasse 48/49
建筑设计师 Architect：	SILVIA CARPARNETO ARCHITEKTEN, FATKÖHL ARCHITEKTEN, BAR ARCHITEKTEN
基地面积 Site Area：	7400 m^2
建筑面积 Total Area：	8880 m^2
项目成立时间 Founding Time：	2010
项目状态 Project Status：	在建 under construction

Fig.3.2项目区位及实景 Project location & scene

紧邻柏林米特区施普奈河畔，伴随着狂野的电子乐声，这里曾经是柏林人彻夜狂欢的场所，也是东西柏林之间的无人看管区，Spreefeld 项目(SFB)正是位于此地(Fig.3.2)。

SFB 项目目前共有 60 个成员，建筑是由三栋住宅综合体所组成，居住空间由于空中连廊的存在，实现了立体化的使用。项目的目标是通过成员合作来整合资源、强化参与并建立自助互助的生活模式。参与的重点放在了设计和创造共享空间上，包括共享空间的使用功能、空间类型以及空间组织模式。

底层空间提供公共、商业、社会、邻里及社区等方面的服务。在这一层面上，项目创造性地在紧靠生活区域的位置设置了占建筑底层面积约 40% 的“选择性空间”(Optional Space)，这是该项目永久保留的功能空间，专为那些拥有先锋城市理念而苦于没有实践场所的人所设置。只要拥有好的想法，通过展示说明说服项目组成员，就可以低廉的价格租用这些场地。通过这种方式引入更多人参与到项目的规划中来，集思广益，最大最好地利用这些共享空间。而这一概念并不是凭空产生的，该场地原本就有作为临时性使用的开放公共使用空间，项目团体只是将其整合到设计之中，以“选择性空间”的形式予以延续。

Close to River Spree with the wild electronic music here used to be the crazy party zone for Berliners and the no-man's land between east and west Berlin. Project Spreefeld Berlin is located here(Fig.3.2).

The Co-housing group consists of 60 members and the project contains 3 housing complexes. With the "sky corridors", residents realize the 3D use of the living space. The objective is to integrate resources, strengthen participation, establish self-help service and to explore an alternative living model under the teamwork of the group. The focus of participation was laid on the design of shared communal spaces including their functions, space types and space Organization patterns.

The underground space is permanently reserved for public, commercial, social uses and other community services. Close to the living area, the 40% of the underground space is used as "Optional Space", which is designed for those urban pioneers who have good practical concepts but cannot find suitable places to realize them. After they persuade the group members successfully with idea

Fig.3.3 Spreefeld Berlin 共享空间体系示意图
Shared Space System of Spreefeld Berlin

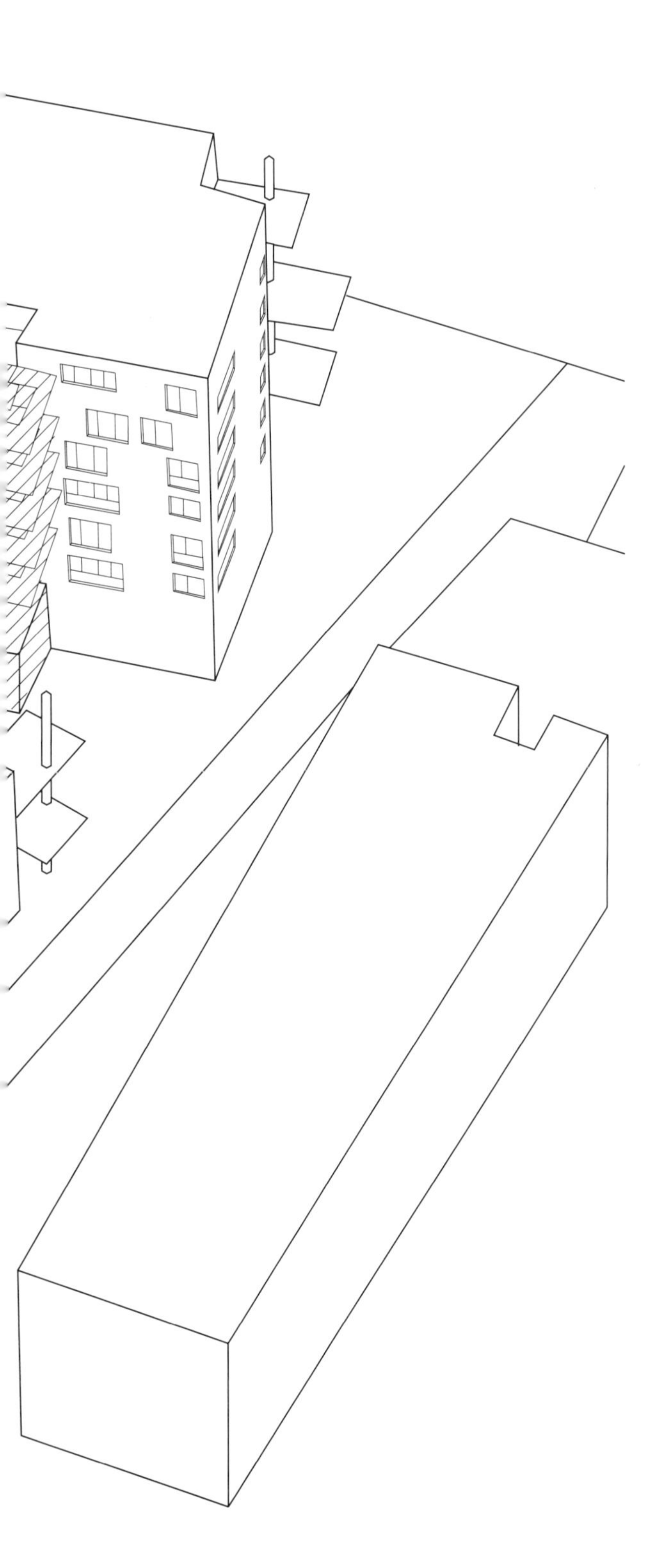

不同于普通住房项目占据空间为已用，联建项目能够从城市角度着手，开放资源给城市。这一案例就将沿河区域完全开放出来，修建了滨水步道，让市民能够亲近施普奈河（River Spree）水域，同时也增加了活动产生的可能性，社区的面貌也由此形成。

室外的共享空间不仅能起到丰富社区生活的作用，同时也起到了联系室内共享空间的作用，在本案中，空中连廊，露台、中央庭院以及滨水步道等室外共享空间整合在一起，配合室内的共享空间，整个空间变得立体同时有层次（形成了不同等级的公共空间），提升了用户对于共享空间的利用效率（Fig.3.3）。

presentation, they can get the “Optional Space” with low rent. With this method, the Co-housing group wants to attract more people to the plan of the project and to make the best and most use of the space. It has be to pointed out that the “Optional Space” idea is not an innovation one but is the heritage of the site itself. The site was used to be the public space and many temporary uses were tested there. The Co-housing group just integrated this former function into the design.

Unlike the normal real estate projects, occupying the space for self-use, Co-housing projects often start from the urban level and offer available resource back to the city. In this case, the riverside space is open to the public and waterfront decks are built for the citizens, thus an active community atmosphere forming.

Shared outdoor space contributes not only to the ample community lives but also to the connection to the shared indoor space. In this case, “Sky corridors”, terraces, central courtyard and the waterfront decks are integrated together. Combined with the shared indoor space, they build up a 3D communal space system and promote the space utilization of the users (Fig.3.3).

02 Zwillinghäuser

街角还给城市
Return the Street Corner Back to the City

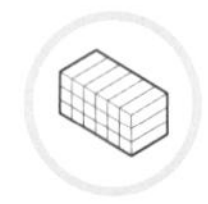
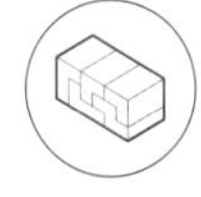
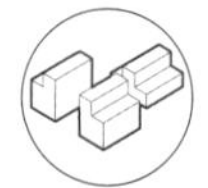

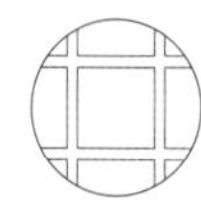

项目概况 Project Profile

项目名称 Project Name： Zwillinghäuser
项目地址 Project Address： Schönholzer Str.10A
建筑设计师 Architect： ZANDERROTH ARCHITEKTEN
基地面积 Site Area： 1466 m^2
建筑面积 Total Area： 4546 m^2
项目成立时间 Founding Time： 2005
项目状态 Project Status： 完工 finished

Fig.3.4项目区位及实景 Project location & scene

Zwillinghäuser 项目位于米特区市郊城市再开发区域的东北部，靠近柏林墙贝尔瑙尔大街段 (Fig.3.4)。该区域的特点是：街块由小尺度分隔的地块所构成，建筑多是在古典晚期和德国经济繁荣时期（Gründerzeit）建成。东西向的贝尔瑙尔大街是一条重要的交通线路，同时也是主要的商业街。

自 2002 年起，柏林政府就开始在这块高密度的区域进行投资，并将开辟新的绿地和开放空间作为重要的发展目标。新建筑及老建筑的修复更新极大地改善了社区的基础设施。

到 2007 年年底，大约 80% 的老建筑已经被翻新改造，此外，许多空地也被新建筑所填充，这其中就包括了本案例。这片区域所发展起来的活力以及不断完善的基础设施吸引越来越多的年轻家庭来此定居，该区域的新建项目中有很多都是针对这一人群而设立的，Zwillinghäuser 项目正是其中的代表，它由两栋多家庭住宅所组成。

项目于 2007 年 4 月竣工，所处地块是由原来分属不同业主（联邦所有、地区所有以及社区所有）的三块用地所组成。不同于最大限度占满用地的惯常做法，该项目选择了其中两块相对的地块分别紧靠既有建筑的山墙面建造了六

Project Zwillinghäuser is located in the northeast part of 38 ha redevelopment zone in Mitte, next to the Bernaustrasse, which is an important traffice line and commercial street(Fig.3.4). The blocks here are divided into small size plots and the buildings were built mainly in the Gründerzeit period.

Since 2002 Berlin Senate has invested in this high-density area and the development focuses on the green and open space. Trees are planted and playgrounds for the children are added and extended. Integrated measures are taken to lower the traffic noise and to improve the safety. New buildings and renovated buildings greatly improve the basic infrastructure in the communities. By 2007 about 80% of the old buildings have been renovated and many vacancies have been filled by the new buildings. More and more populations are attracted to move here especially many young families. Under this background many projects are initiated and among them is the project Zwillinghäuser. It consists of two multi-generation houses and it was finished in 2007.

The site is made up of three lots belonging to different

Fig.3.5实景 Scene

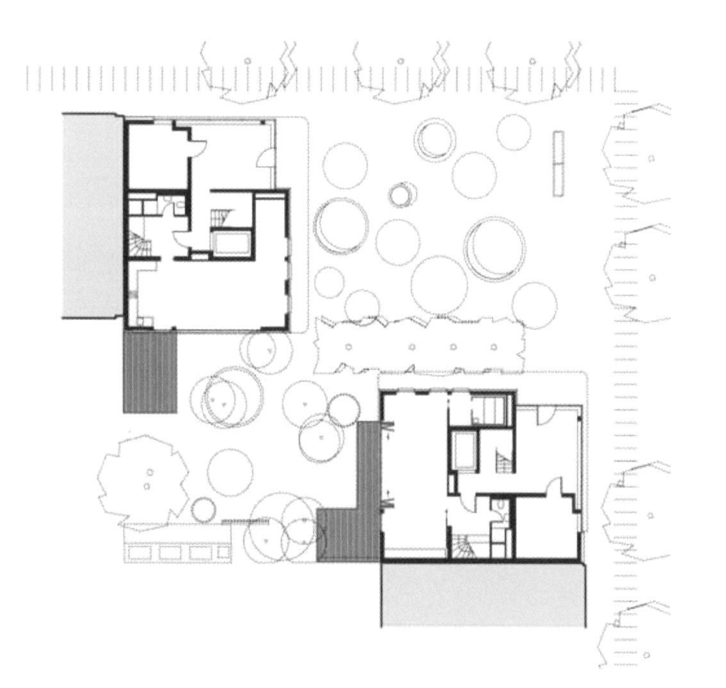

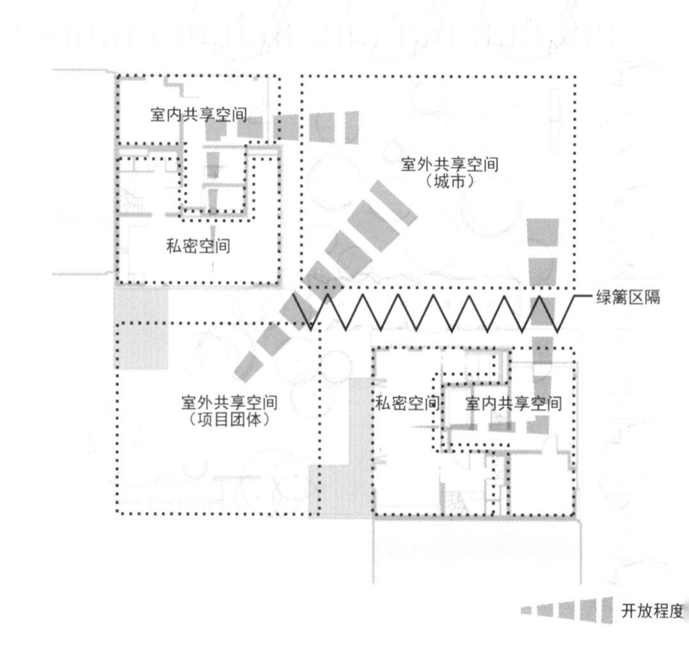

Fig.3.6 共享空间体系示意图 Shared communal space system

层高的住宅，从而空出了占据街角的第三个地块。这样做，一方面使得住宅能够获得更多的日照；另一方面，开放的街角成为了城市的公共广场。这个街角广场通过设置如座椅等城市家具来吸引人们在此休憩驻足，为区域周边提供了对外开放的城市共享空间，提升了这片人口密集区的城市空间品质，并且团体保证，在接下来的 100 年里他们将负责整个广场的维护和管理工作。

与开放的街角广场相对应的，是住户所保有的私人庭院。为了给住户家的孩子提供安全的游乐场地，整个庭院由篱笆环绕，并且始终处在各户视线的监控范围内。除了私人庭院外，包括街角广场在内的共享空间都种上了树木，为夏日居民的游乐以及交流提供了树荫（Fig.3.5）。

在本案中，外部共享空间是存在一定的序列和等级的。开放程度最高的是街角广场，它面向城市开放，通过绿篱与内部共享庭院进行区隔，但视线仍可穿透，只是对开放权限进行了缩减。同时室内的共享空间（同样是限定在项目内部共享使用）连接了室外共享空间和私密空间，起到了空间上的缓冲作用（Fig.3.6）。

owners (federation-owned, district-owned and community-owned) which has less attraction to the general investors, so the group purchased them at a relatively low price. Instead of occupying all the site, this case has chosen the two opposite lots to build two 6-story housing which is next to the gable of the existing buildings, thus a corner square is reversed. On one hand through this way the housing can get better sunlight, on the other hand it offers a public square to the neighborhood. The Co-housing group promises 100 -year maintenance and management to the square.

This project also has a private inner courtyard which is only open to the members. This space is surrounded by hedges, providing children with a safe playground. Trees are planted in the square and benches are also set there, offering people shades in Summer and meeting points every day(Fig.3.5).

In this project, shared outdoor spaces are organized according to the designed space sequence and open hierarchy. The most public space is the street square which is open to the city. There are hedges between the inner courtyard and the square which can ensure the visual connection between the two parts while the degree of openness reduces in the inner yard. Besides the shared indoor space on the ground floor (shared only by the residents) links the outdoor space and the private space, acting as the space buffer (Fig.3.6).

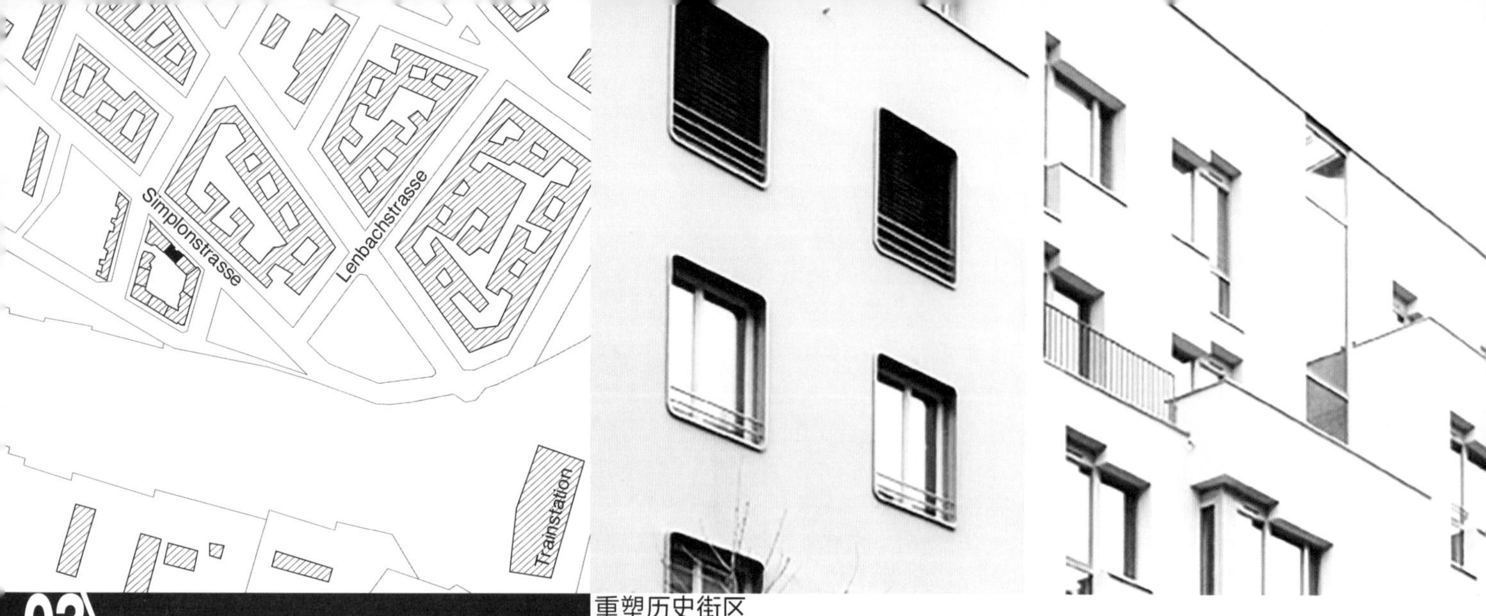

03\ Baugemeinschaft Simplon

重塑历史街区
Rebuild the Historic Block

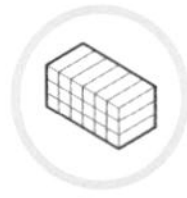  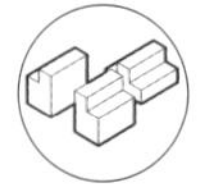

 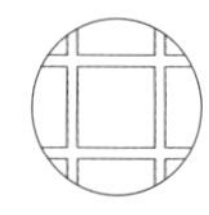 

项目概况 Project Profile

项目名称 Project Name：	Baugemeinschaft Simplon
项目地址 Project Address：	Simplonstrasse 54
建筑设计师 Architect：	FATKÖHL ARCHITEKTEN
基地面积 Site Area：	N/A
建筑面积 Total Area：	1828 m^2
项目成立时间 Founding Time：	2008
项目状态 Project Status：	完工 finished

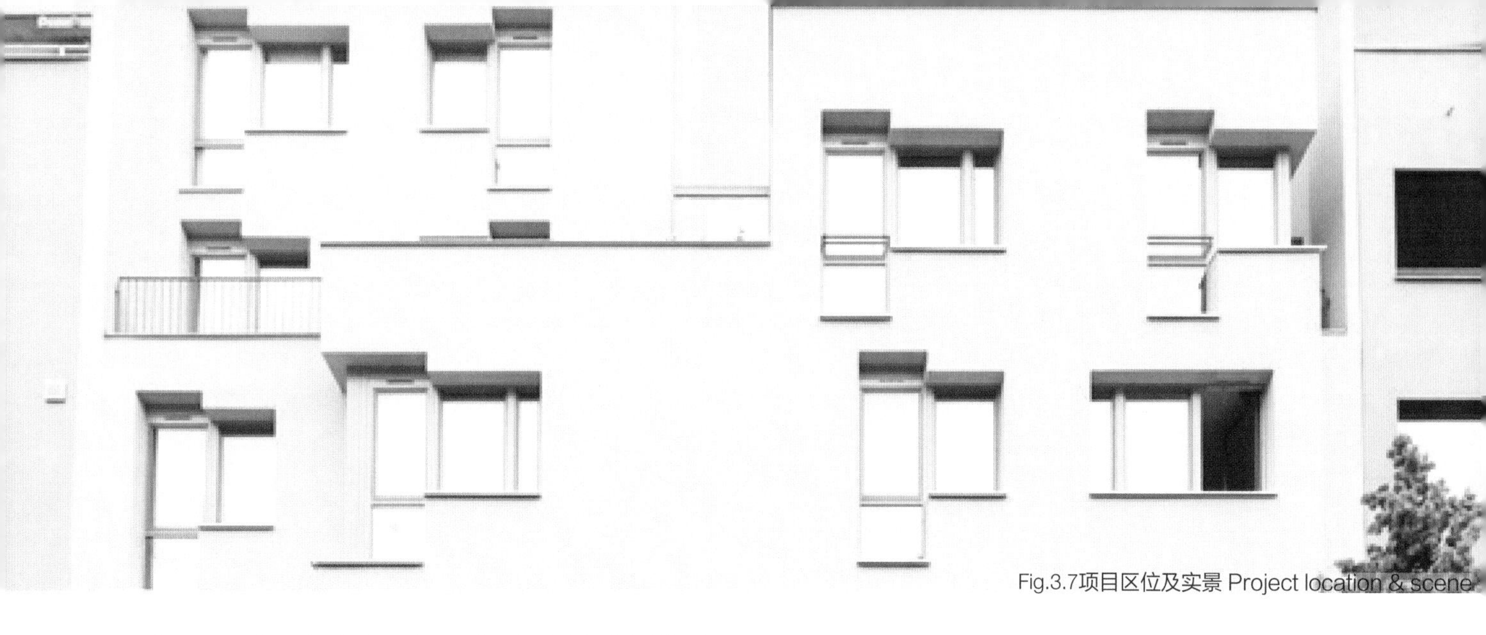

Fig.3.7项目区位及实景 Project location & scene

Baugemeinschaft Simplon 项目位于柏林弗里德里希，这是一个由多个项目联合而成的综合性项目，整个街块只有一栋 1930 年代的老建筑遗留下来。项目团体找到了这一地块并开始实践 (Fig.3.7)。

街块按照历史的边界进行划分，这样得到的正常大小的场地被分配给了彼此独立的联建住宅项目团体，建设包括 A,B,C,D,E,F,G,H 八栋建筑 (Fig.3.8)，本案例是建筑 C。共享的屋顶露台提供了观赏柏林城市景观的绝佳位置。到 2013 年底，整个街块顺利闭合，所形成的内部庭院由这几个联建项目团体所共享。此外，建筑地下停车库与游乐场地也由社区共享。内部庭院与外部街道不在同一标高上，并有开口通向街道，在空间上有某种“进入”的暗示，同时也有项目在建筑内部提供了连接内外空间的廊道。

在这里，共享空间存在等级：廊道由本栋建筑成员所共享，并设置了自行车库以及连接内部交通核的通路。廊道连接两端，一端为城市层面的街道空间，另一端则为共享的庭院。在面向街道的一侧，设置的商业办公空间进一步缓冲了城市界面对私密空间的干扰 (Fig.3.9)。

Project Simplon is located in Berlin Friedrichshain and consists of 8 Co-housing groups(Fig.3.7). The block was almost damaged in the WWII. Based on the historical block division, 8 lots are distributed to 8 groups and thus the project contains 8 buildings(Fig.3.8). The whole block was not finished until 2013. The inner courtyard is shared by the groups and the shared roof terrace provides a perfect place to overlook the city. Besides, all the Co-housing groups share a underground garage. The inner courtyard is lower than the street outside and there is a big ramp exit connecting the two parts while in each building there is an inner linking passage.

There is also hierarchy in the shared space.The linking passages are shared within the buildings and along them bike garages and access to the staircase are arranged. One side of the passages is the street space in the urban level and the other side of the passages is the shared inner courtyard in the community level. Meanwhile some commercial spaces and offices are arranged in the street side to prevent the private space from the disturbance of the urban interface (Fig.3.9).

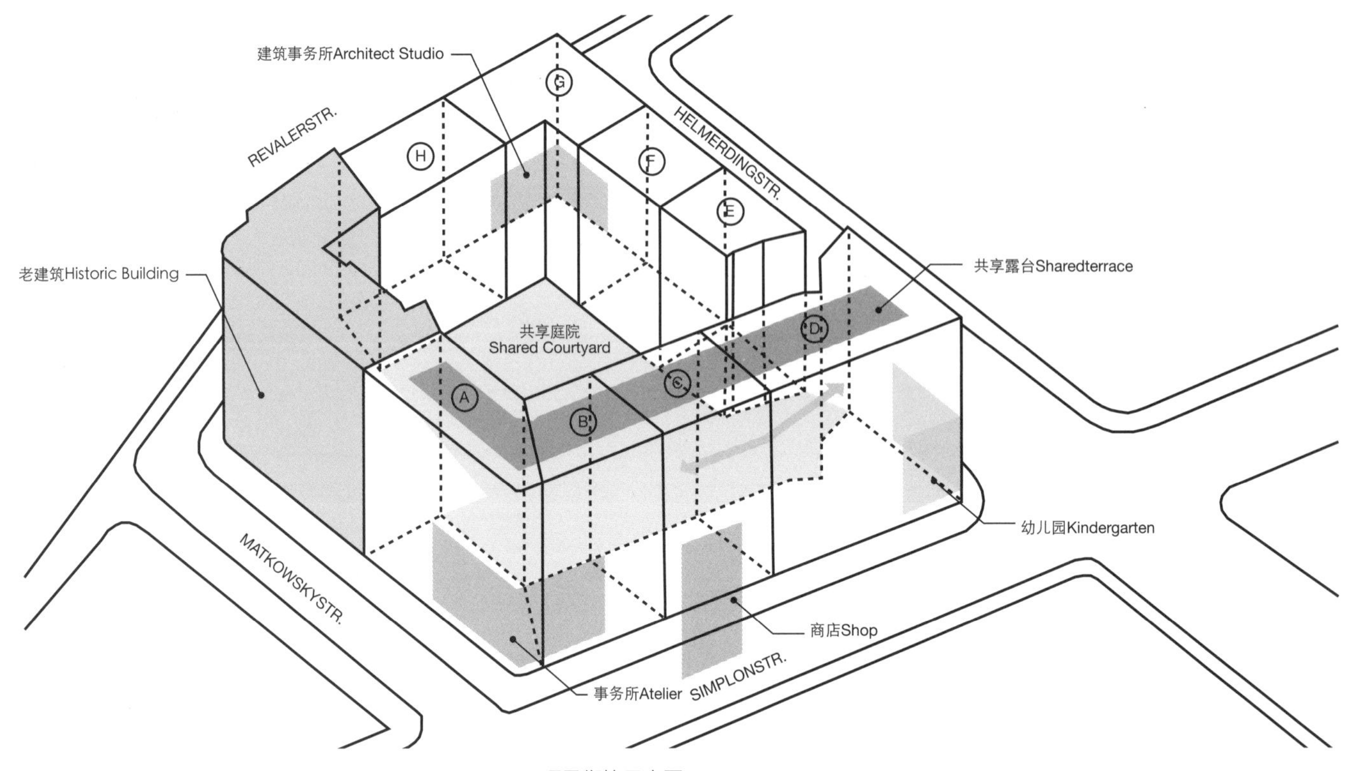

Fig.3.8项目街块示意图 Block Demonstration

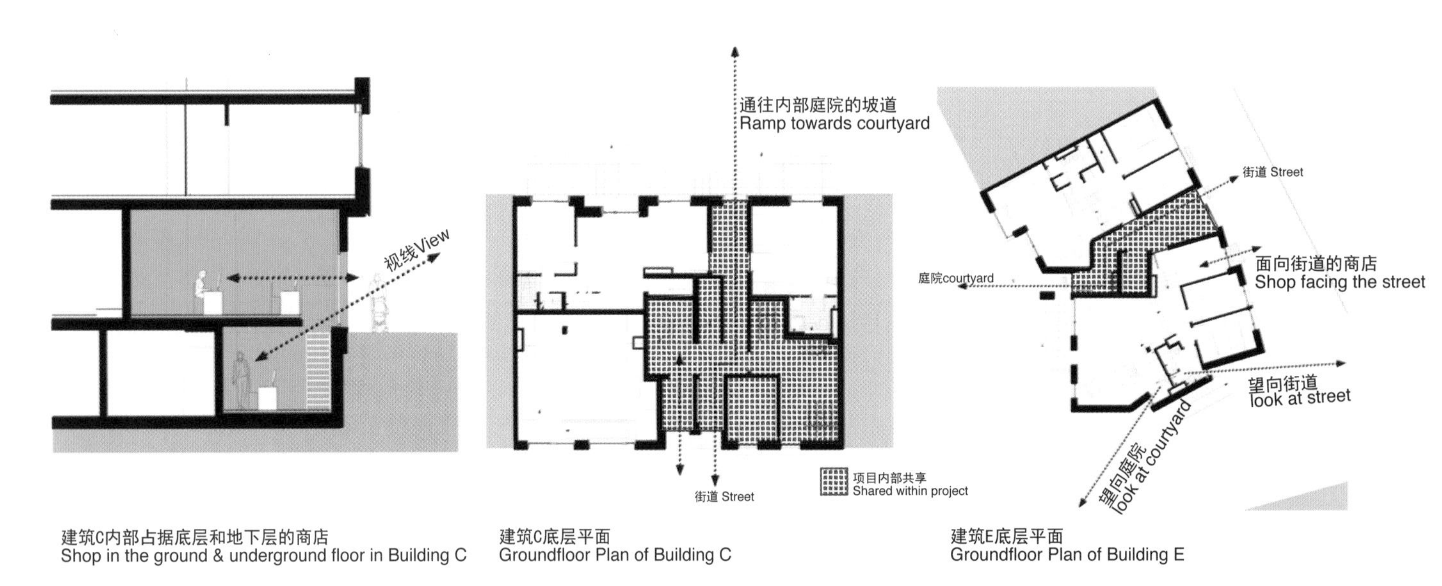

Fig.3.9共享空间分析 Shared Space Analysis

3.1.2 内向的参与：共享的室内空间

长期公平平等的讨论交流，最终形成了体现集体意志的、几乎在所有联建住宅项目中占据重要地位的共享空间。共享空间就其共享性存在着不同等级，从城市范围的共享，到街块范围，再到建筑内部，共享的人群范围逐渐缩小，功能涵盖更多，空间形式也更多。如果说共享的室外空间在普通住宅项目中也十分常见，甚至被作为市场噱头来进行宣传，那么共享的室内空间则使得联建住宅项目显得更加独特。

3.1.2 Inward Participation: Shared Indoor Space

The long-term democratic discussion and communication finally give birth to the shared communal space which embodies the collective will and it occupies the important position of the Co-housing. Share-ability hierarchy exists in the shared space: from the urban level, to the community level, and to the building level, the group range that shares the space becomes narrower and narrower, the function becomes more and more multiple and the space form becomes more and more. Usually shared outdoor spaces also exist in the normal real estate projects sometimes even as the promotion boasts, but what really distinguish them from the Co-housing are the shared indoor spaces.

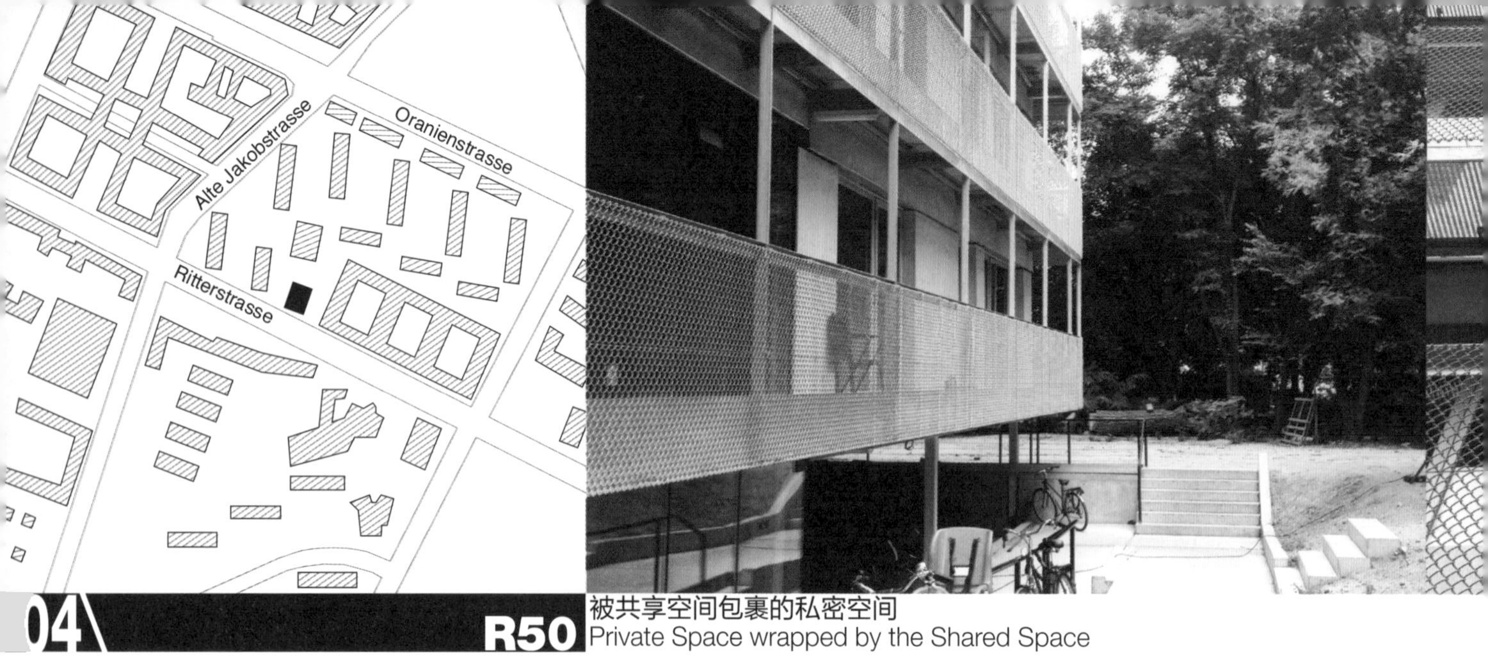

04 R50 被共享空间包裹的私密空间 Private Space wrapped by the Shared Space

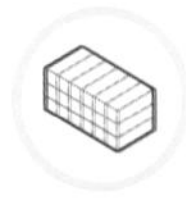 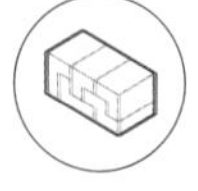 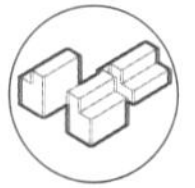

 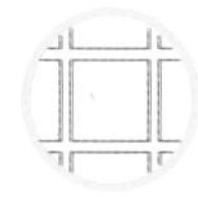 

项目概况 Project Profile

项目名称 Project Name：	R50
项目地址 Project Address：	Ritterstrasse 50
建筑设计师 Architect：	IFAU+JESKO FEZER　HEIDE & VON BECKERATH ARCH
基地面积 Site Area：	2055 m^2
建筑面积 Total Area：	2780 m^2
项目成立时间 Founding Time：	2010
项目状态 Project Status：	完工 finished

Fig.3.10项目区位及实景 Project location & scene

R50 项目位于里特大街 50 号，项目团体由 20 户人家组成，他们大多是富有创意的中产阶层年轻教授，而项目的目标是建造一栋尽可能节能的住宅。项目中最关键的部分是在轻轨环线内提供可负担的居住空间并增强彼此之间的交流(Fig.3.10)。

本项目是一栋独立公寓，由未来入住的居民共同完成设计。建筑师在设计初期负责完成建筑空间的基本分配，包括交通流线、基础设施以及主体空间框架，剩余的包括共享空间以及公寓平面则由居民参与完成。

在 R50 项目中，共享空间扮演了十分重要的角色。大型共享空间几乎占据了整个底层。此外，大家还共享拥有露天厨房的屋顶露台以及围绕在建筑每层的外廊空间。该项目更具深层意义的是其集体设计的过程，可大致描述如下：首先通过两次会议确定是否需要共享空间，需要哪些共享空间以及每种共享空间的资金筹措方式；然后通过十次会议确定共享空间的位置：是在屋顶，还是分布在每一层抑或是设置在底层；然后再讨论共享空间的空间尺寸，居民根据不同的共享功能分别写出其理想面积；然后设计师统计其最大值与最小值，再取其平均值；最后讨论应该如何使用共享空间，最终形成了如下所述的共享空间格局：

Project R50 is located in Ritterstrasse 20. The Co-housing group is composed by 20 parties. Most of them are creative, middle class, and young professionals. The objective was "to build a house as cost-efficiently as possible". One important thing was to enhance community and another important thing was to provide affordable living space within the S-Bahn ring(Fig.3.10).

"It is a freestanding building and was designed in a participative scheme in which residents were involved. The architects proposed at the beginning basic spatial distribution of accessibility, infrastructure and grid structure. The rest of the house, including common areas and apartments ground plan was designed with the participation of the dwellers.[1] "(Corzo, Arredondo, 2013)

In R50 shared spaces play quite important roles. Large shared space is well equipped with facilities, guest rooms and the path towards the garden takes the whole of the underground floor. Besides they share a roof terrace with summer kitchen and the corridor space around each floor.

1 Source: Corzo, D., Arredondo, N. (2013) Urban Commons, unpublished thesis (M.A.), TU Berlin

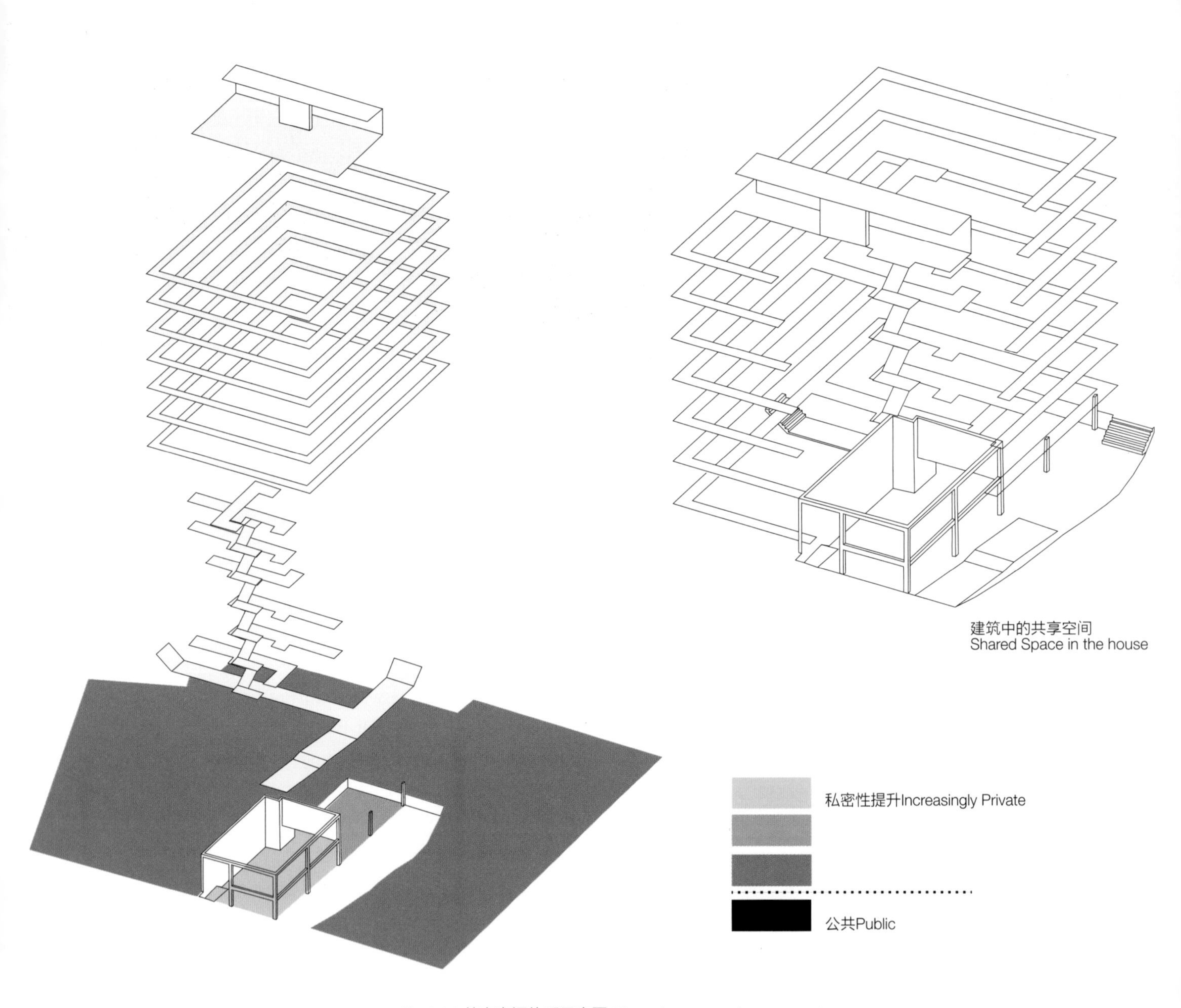

Fig.3.11 共享空间体系示意图 Shared communal space system

共享性最强的空间位于首层，这里也是每户居民使用最频繁的场地，包括多功能室、自行车库等，其次是环绕各层的外廊空间，起到了扩充室内储藏空间的作用，使用的人群则主要集中在同层住户，还有屋顶露台，其使用频率则不及前两类空间（Fig.3.11）。

What is more meaningful is the progress of the collective design. They were asked to describe the communal space including its function, size and position. Then the architects collected the information, evaluated them and made a full consideration.Finally the system of the shared spaces formed: the most used shared space-multifunctional room is in the underground floor. Then are the shared corridor spaces on each floor ,which are often shared by the residents on the same floor, acting as the extension of the inner space; The roof terrace with the summer kitchen is often used in Summer. The Shareability Hierarchy is showed in the Fig.3.11 .

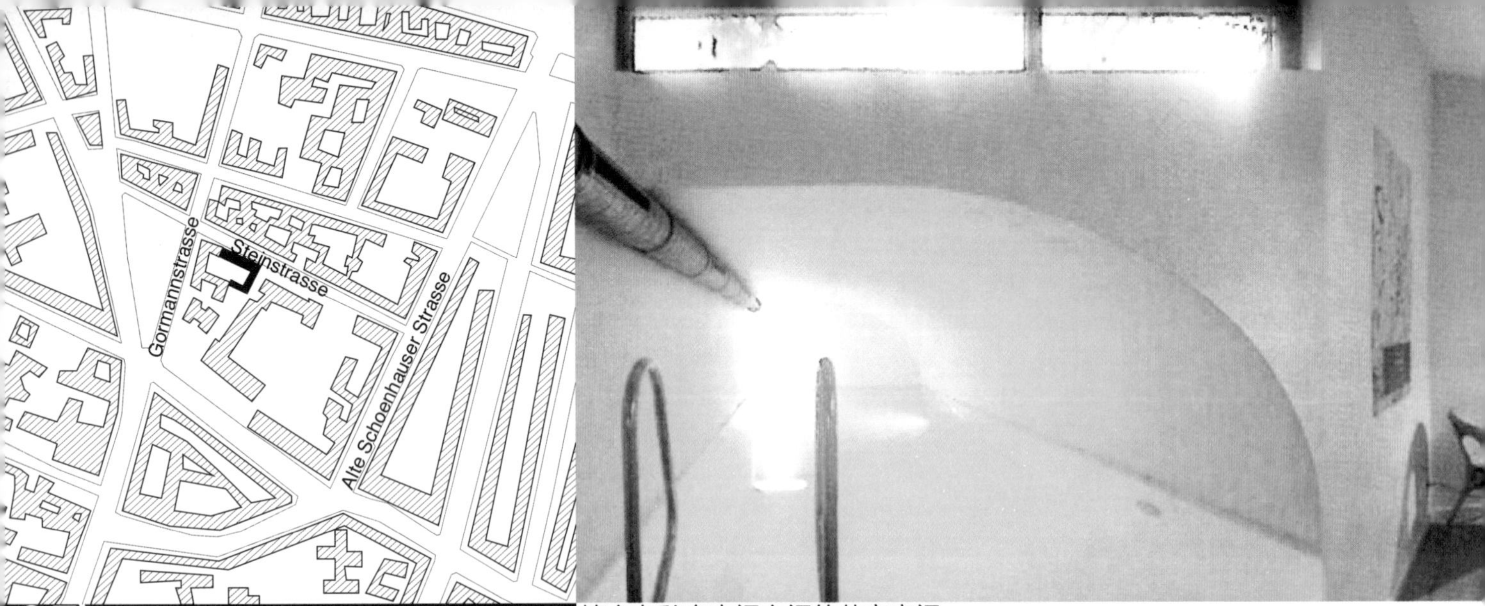

05\Wohnetagen Steinstrasse

填充在私密空间之间的共享空间
Shared Space inserted in the Private Space

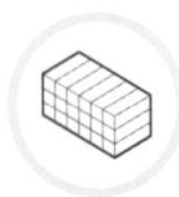 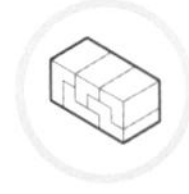 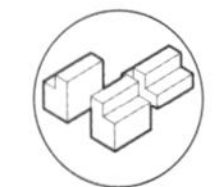

 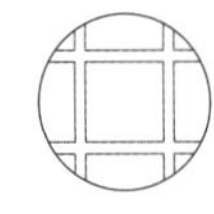 

项目概况 Project Profile

项目名称 Project Name： Wohnetagen Steinstrasse
项目地址 Project Address： Steinstrasse 26-28
建筑设计师 Architect： CARPANETO SCHÖNINGH ARCHITEKTEN
基地面积 Site Area： 1500 m^2
建筑面积 Total Area： 3400 m^2
项目成立时间 Founding Time： 1998
项目状态 Project Status： 完工 finished

Fig.3.12项目区位及实景 Project location & scene

坐落于热闹的柏林米特区中安静的施坦大街，该项目拥有 22 间公寓和 5 间工作室，它们由各不相同的 $60m^2$ 至 $250m^2$ 的模块组成。建筑师将这个项目描述为大型立体拼图游戏，建筑由不同的居住模块水平向与垂直向地构成。Wohnetagen Steinstrasse 项目在拓展共享空间的功能与形式方面作出了许多探索，这些都是通过讨论以集体的形式确定的。该项目不仅共享了一个停车库，还有花园、露天厨房、工作坊、桑拿房、游泳池、客房以及聚会厅（见图 3.12）。如此丰富的共享资源，为打造一个活跃的邻里氛围做出了不小的贡献。该项目特别的地方在于，住宅项目本身的法律形式是业主委员会（GbR），而共享空间部分则另外成立了有限公司（GmbH）。这样的好处是，对于游泳池等类型的共享空间没有兴趣的住户则可不参与投资。

Project Wohnetagen Steinstrasse is located in Steinstrasse and it has 22 apartments and 5 studios. The buildings consists of modules from 60 m^2 to 250 m^2 and the designer described this project as a Lego-game. It has explored how to extend the functions and forms of the shared space. The group shares a garden, a summer kitchen, a workshop, a sauna room, a swimming pool, guest rooms and an assembly room (See Fig. 3.12). The special point of this project is that the legal form of the Co-housing is GbR while the shared spaces is GmbH. The benefit is that residents show no interest in the swimming pool and other shared spaces don't have to be invested with money.

空间环境的质量
Quality of the physical enviroment

	差 bad	好 good
必要性活动 Necessary Activity		
自发性活动 Optional Activity		
社会性活动 Social Activity		

触发活动的可能性
Possibility of relevant activity

Fig.3.13户外活动发生频率与空间质量的关系
Relation between possibilities of activities and space qualities

传统住宅
室内外空间分隔明显
Traditional Housing
Clear Boundary between Indoor & Outdoor

联建住宅
室内外分隔柔化
Co-Housing
Soft Boundary between Indoor & Outdoor

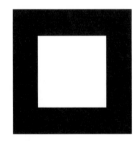

传统住宅
几无室内共享空间
Traditional Housing
Hardly Indoor shared space can be found

联建住宅
室内分布共享空间
Co-Housing
Indoor shared space can be found

室外共享空间
Shared Outdoor Space

室内共享空间
Shared Indoor Space

Fig.3.14参与性与共享性 Participation & Sharability

3.1.3 小结

杨•盖尔在《交往与空间》一书中将户外活动分为三大类：必要性活动、自发性活动和社会性活动。“当户外空间的质量不理想时，就只能发生必要性活动。……当户外环境质量好时，自发性活动的频率增加。与此同时，随着自发性活动水平的提高，社会性活动的频率也会稳定增长”[1]（Fig.3.13），而邻里互动则属于后两类户外活动的范畴，正是通过设置共享空间，为邻里提供了集会场所，整个邻里受益于此，促进了互动、交流，也塑造了较强的社区归属感。开放私人绿地及其他场所给公众需要勇气以及对彼此的信任。从一些实例中作者了解到，这些项目甚至影响了周边的邻里，使得他们也在考虑效仿，这正是由共享空间帮助建立起的信任和理解所促成的。

联建住宅项目成员希望改变现代居住小区中冰冷的人际关系，他们创造了许多效果喜人的生活氛围，而这其中共享空间的作用功不可没，共享空间促进了社交意识的增长及与邻里的互动。联建项目在内城中创造了一个又一个的“城市乌托邦”，而这些项目以传统方式几乎是不可能实现的。在某种层面上来讲，共享空间的规划数量是显示项目潜在社会福利的有效指标，也是塑造区域归属感的空间基础。

在传统的街区中，我们也能找到诸如花园、庭院之类的公共空间，但它们与私密空间之间的界限是十分明确的，而在联建住宅项目里，不同等级的共享空间从室外延续到室内，最后连接着私密空间，二者之间的界限被柔化，并得到缓冲。同时在室内单元中也出现了共享空间，这也是在传统住宅中极为少见的（Fig.3.14）。

在住宅项目中置入共享空间需要注意的问题主要集中在如何解决共享空间与私密空间的关系上，住宅项目归根结底其核心功能还是居住，这涉及到每个人生活中最私密的部分，即便是倡导集体生活的联建住宅项目也无例外。通过对已搜集案例的归纳整理，共享空间与私密空间的空间关系可分为两部分进行讨论，一是室外共享空间与建筑体的关系，

1 扬•盖尔著，何人可译，交往与空间，北京：中国建筑工业出版社，2002

3.1.3 Summary

“Greatly simplified, outdoor activities in public spaces can be divided into three categories, each of which places very different demands on the physical environment: necessary activities, optional activities, and social activities.... These activities are especially dependent on exterior physical conditions.When outdoor areas are of poor quality, only strictly necessary activities occur.When outdoor areas are of high quality, necessary activities take place with approximately the same frequency- though they clearly tend to take a longer time, because the physical conditions are better. In addition, however, a wide range of optional activities will also occur because place and situation now invite people to stop, sit, eat, play, and so on. [1]”(Jan,1987) (Fig.3.13) Neighborhood Interaction contains the latter two categories: optional activity and social activity. By offering shared space for the neighborhood, the frequency of such activities will increase and the harmonious community atmosphere will form. From the selected cases, we can find that the contribution made by the Co-housing projects sets good examples for the neighborhood and they also start to offer shared space as Co-housing projects did.

Preliminary Co-housing groups tended to choose the long-term ignored plots or noman’s lands to practice the projects. Most of them want to change the cold social relationship existing in the modern living communities. Community-oriented Co-housing projects have created ideal living atmosphere and warm relationship among the residents has been established. Shared spaces contribute a lot to it and promote the social awareness and neighborhood interaction of the residents. By this way one after another Co-housing creates “Urban Utopia” in the inner-city, which normal real

1 Source: Jan, G. (1987) Life between Buildings, New York: Van Nostrand Reinhold Company Inc.: 11-13

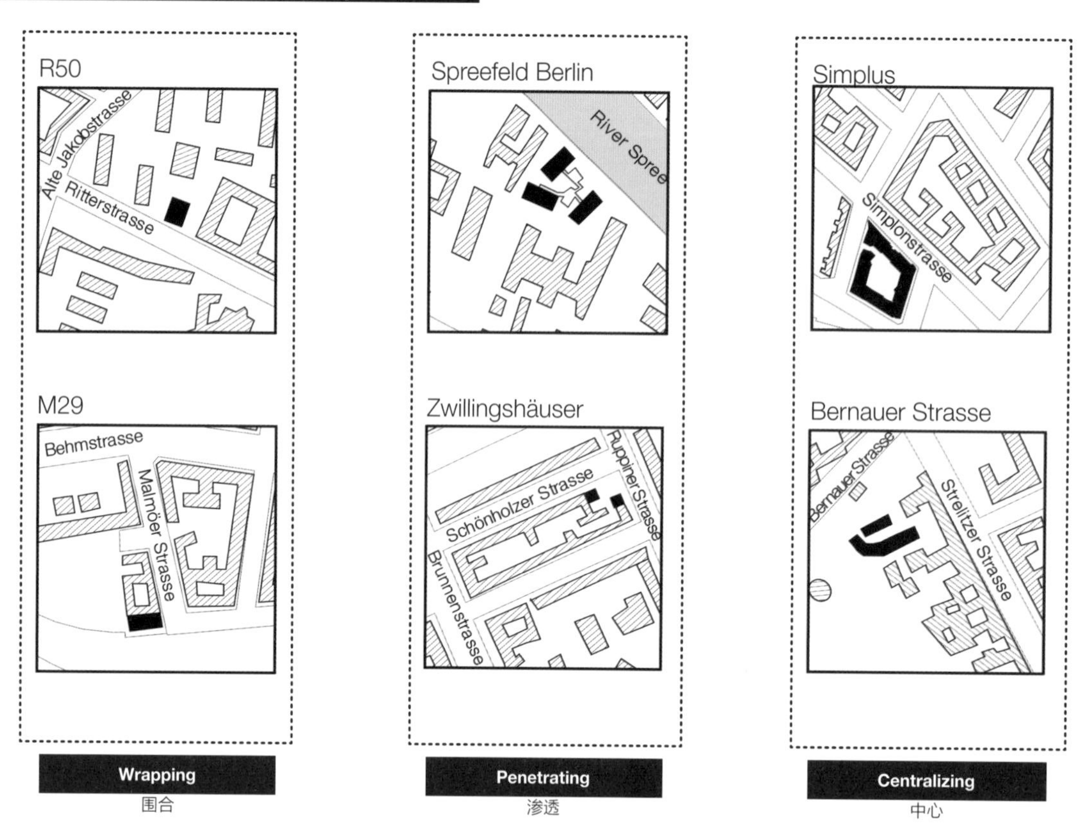

共享的室外空间和建筑之间的空间关系
Relationship between shared outdoor space and the buildings

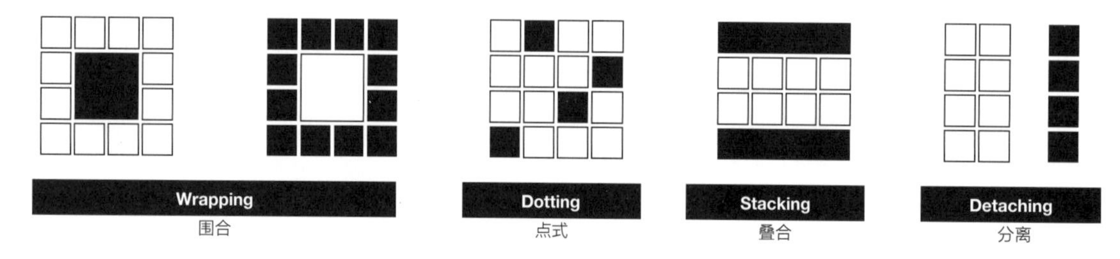

共享的半室外空间、室内空间和私密空间之间的空间关系
Relationship between shared semi-outdoor space, shared indoor space and the private space

Fig.3.15不同性质空间之间的空间关系 Relationship between different spaces

一是室内及半室外共享空间与私密空间的关系。

室外共享空间与建筑体的关系按照室外空间位置分布主要有三大类型 (Fig.3.15)：

（1）**环绕型**：建筑体位于场地中心，周边环绕共享空间，开放性较大，一般直接与城市界面相连通。

（2）**渗透型**：建筑体散布，室外共享空间从城市界面一直渗透到社区内部。共享空间存在等级分化，共享人群逐层限制。

（3）**中央集中型**：建筑体为周边环绕式，室外共享空间居于中心位置，主要表现为中央庭院或花园，无形中限制了共享人群的范围。

室内及半室外共享空间与私密空间的关系主要有四大类型（Fig.3.15）：

（1）**环绕型**：这种类型有两种情况，一种情况是共享空间以外廊形式环绕各层，作为私密空间的扩展，一般说来共享级别较低，主要由同层住户所共享，作为对主共享空间与私密空间的过渡区域辅助存在；另一种情况则是共享空间居于中心位置，私密空间环绕布置，一般来说共享级别较高，使用效率也较高，共享空间的地位高于私密空间。

（2）**散布型**：共享空间散布各层，一般说来也是作为次一级的共享空间辅助存在，主要服务于同层住户，共享空间与私密空间趋向同质，这种形式一般适用于酒店型或宿舍型联建住宅项目。

（3）**叠合型**：这种类型颇为常见，共享空间与私密空间相互叠合，共享空间占据首层或 / 和顶层，通常面积较为宽裕功能较为复合，私密空间位于中间层，这种方式使得二者都能正常独立运行，效果较佳。

（4）**分离型**：这种类型较少见，共享空间完全脱离私密空间独立存在，二者皆能正常独立运行，见于空间资源较富余的项目。

estate projects can hardly achieve. "*In this way, the planned amount of shared space can be good indicator of the potential the project has for adding to society* [2]"(Ring,2013)

In normal real estate housing projects we can also find gardens, courtyards and other public space, but the boundary between the public and the private is clear and definite while in Co-housing projects, space of different shareability degree last from outdoor communal space to indoor communal space and finally connects to the private space. The boundary between the public and the private is soften. Even among the living units, shared space can also be found which is seldom done in normal housing. (Fig.3.14)

That how to insert shared space in the housing projects focuses on how to deal with the relationship between the shared space and the private space. The core function of housing is dwelling which is the most private part of personal lives, even in Co-housing. Based on the selected cases, the discussion on the relationship between shared space and private space is divided into two parts: one is the relationship between shared outdoor space and the buildings; the other is the relationship between shared semi-outdoor space, shared indoor space and the private space.

Relationship between shared outdoor space and the buildings can be summarized as 3 categories (Fig.3.15):

(1) **Wrapping Type**: the buildings are located in the central part of the site wrapped by shared outdoor space. This type has large openness and direct connection to the urban interface.

(2) **Penetrating Type**: the buildings are scattered in the site. Shared space continues from the urban interface to the community and the buildings. There is a clear hierarchy in

2 Source: Ring, K. (2013) Selfmade City Berlin: Stadtgestaltung und Wohnprojekte in Eigeninititie, Berlin: JOVIS: 31

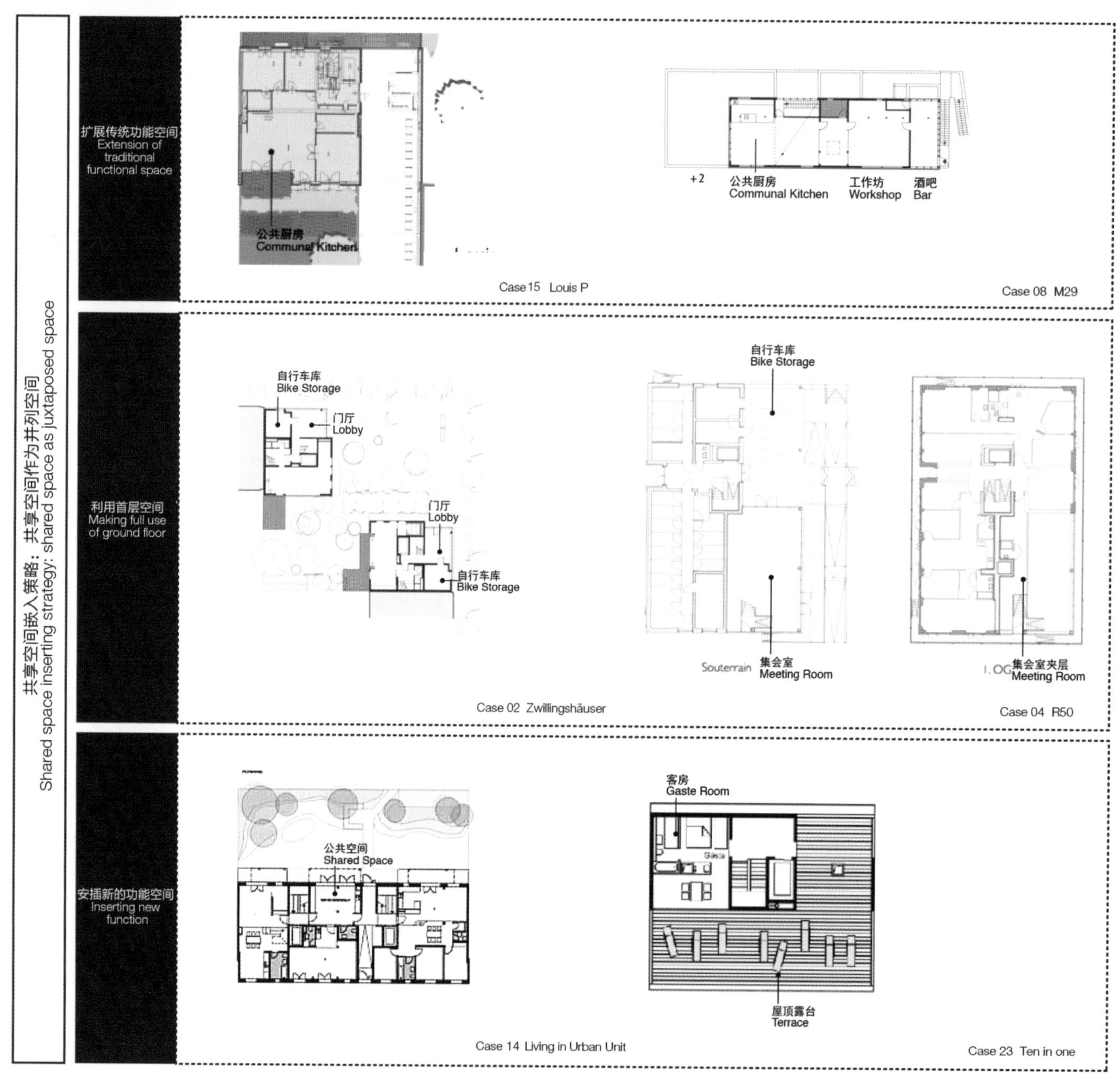

Fig.3.16共享空间嵌入手法：作为并列空间 Ways of inserting shared space: as juxtaposed space

在公共建筑中，人们对于共享空间司空见惯，甚至近年来在一些学生公寓中也引入了这种类型空间，如斯蒂文•霍尔设计的麻省理工学院学生宿舍，JDS 建筑事务所设计的位于法国巴黎的学生公寓“M6B1"。甚至在北京当代 MOMA 住宅楼中，霍尔也引入了共享空间，不过笔者走访之后发现其使用状况不佳，并没有实现真正意义上的共享，所谓的共享空间更多的是来自于建筑师的美好幻想。而在联建住宅项目中，共享空间真正地与住宅相结合，与居民的切实需求相结合，可以说是创造了一种新的住宅类型。下面将总结在住宅空间中嵌入共享空间的主要手法，主要分两个层面：

（1）**共享空间作为并列空间**（Fig.3.16）：

作为与私密空间并列的共享空间，它不单纯依靠扩充辅助性空间来实现，而是分担了传统住宅中的重要的功能空间，并且创造了一些传统住宅空间中不曾有过的功能空间。主要的手法有：扩展传统功能空间（如厨房、餐厅等），将原来散布于各户的这些空间集中到一起，并且复合多重功能；利用首层空间，将室外共享空间延续至室内，安排一些集会空间，车库等；安插新的功能空间如作坊、酒吧、聚会厅、客房等。

（2）**共享空间作为从属空间**（Fig.3.17）：

作为私密空间的从属部分，共享空间通过扩充既有的、存在于传统空间中的功能空间，复合功能实现共享。主要的手法有通过扩展交通空间等辅助性空间来实现，如扩大门厅、楼梯间的休息平台等；利用屋顶平台、露台、阳台等半室外空间。

the shared spaces and the groups that can share the space are defined.

(3) **Centralizing Type**: the buildings are organized in the traditional block way. Shared outdoor space is centralized as the central garden or courtyard. The space is well defined sometimes as well as the groups that can share the space.

Relationship between shared semi-outdoor space, shared indoor space and the private space can be summarized as 4 categories (Fig.3.15):

(1) **Wrapping Type**: this type has two situations: one is that shared space wrap the private space on each floor like the form of exterior corridors. It is used as the extension of the private space and shared by the residents in the same floor. Usually in this situation it has a lower shareability degree and the shared corridors act as the buffer area between the major shared outdoor space and the private space; the other is that the shared space stands in the central position and is wrapped by the private space. Usually in this situation it has a higher shareability degree and higher utilization.

(2) **Dotting Type**: this type of shared spaces is distributed on each floor and is used as secondary shared space majorly serving the residents in the same floor. Shared space and private space become homogeneous. This type is used in the Hotel pattern or dormitory pattern of Co-housing.

(3) **Stacking Type**: shared space and private space stack together. Usually shared space occupies the ground floor or/ and top floor (roof terrace) and has sufficient areas and multiple functions; private spaces are on the middle floors. By this way the two parts can operate separately and well.

(4) **Detaching Type**: this type is rare. Shared space and private space are detached, forming two independent buildings and operating separately. Usually it is used in the

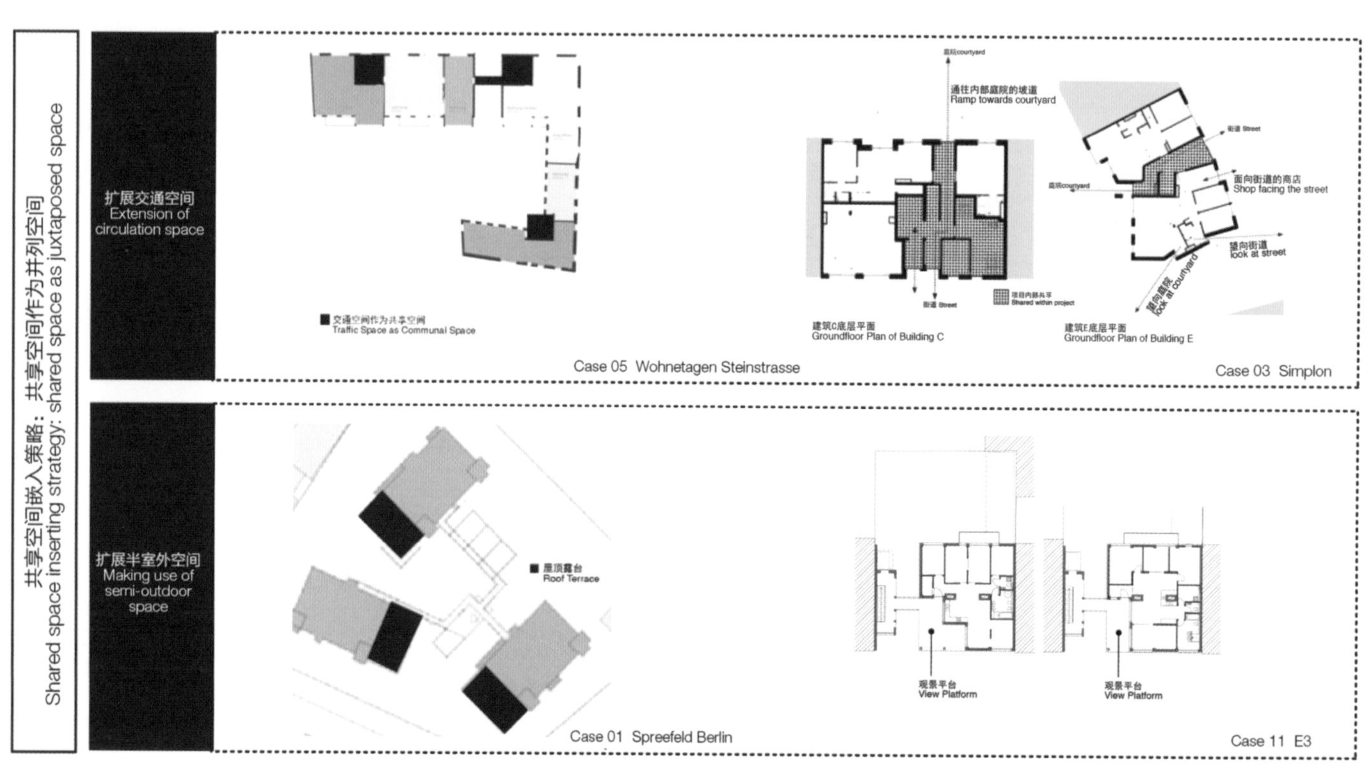

Fig.3.17共享空间嵌入手法：作为从属空间 Ways of inserting shared space: as subordinate space

projects with ample space resources.

In public buildings people used to the shared space. In recent ten years shared spaces are even introduced to the housing projects like Steven Holl's housing project: Linked Hybrid in Beijing. The author visited there but found the shared space there were not well used and they are an ideal concept of the designer himself. In Co-housing projects the shared space embodies the willing of the residents and it realizes the real combination of the shared space and the private space. That how to insert shared space into the private space will be discussed in two levels:

(1) Shared space serving as juxtaposed space (Fig.3.16):

As to juxtaposed space serving as the private space, this type of shared space shares the important functional space in the traditional living units and at the same time it creates some new functional space that never exists in the traditional housing. Major methods can be summarized as follow: extension of the traditional functional space (kitchen, dining room and so on); making full use of groud floor and extending the shared outdoor space to the inner space, arranging functions like assembly rooms and garages;Inserting new functions (workshops, bars, guestrooms and so on).

(2) Shared space serving as subordinate space (Fig.3.17):

As to subordinate space serving as to the private space, this type of shared space makes full use of the functional space existing in the traditional housing, extends them and puts multi-function on them. Major methods can be summarized as follows: extension of the traffic space (lobby, staircase and so on); making use of semi-outdoor space (roof terrace, balcony and so on）.

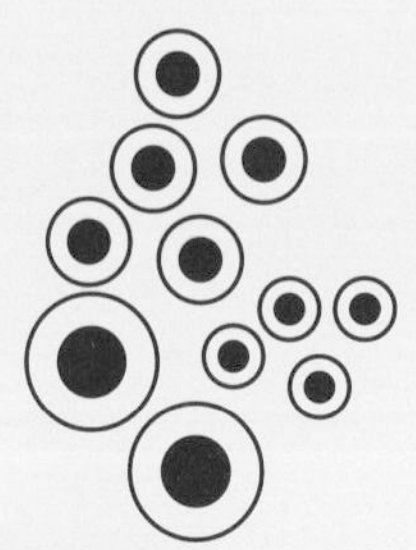

传统住宅
"孤岛"：户与户之间交流甚少

Traditional Housing
Isolated isles:little communication between neighbours

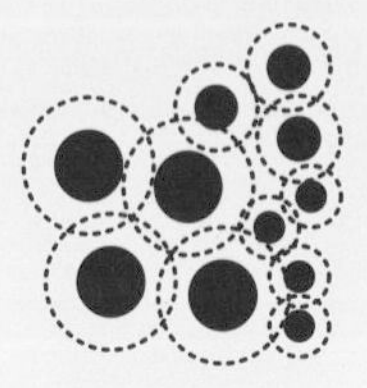

联建住宅
社区活动频繁

Co-Housing
Strong communication between neighbours

传统住宅
几种标准单元

Traditional Housing
Several standard units

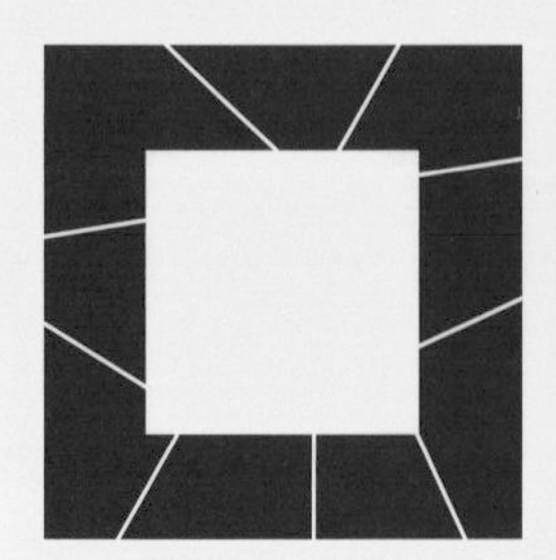

联建住宅
多种单元形式

Co-Housing
Various unit forms

Customization → Identity

Customization
订制性

need to customize their private space meeting their personal living requirements and enjoy personalized life.
根据个人需求订制生活空间，享受个性化生活

Identity
个性

Traditional Housing

Isolated isles:little communication between neighbours / Several limited standard units

Co-Housing

Strong communication between neighbours / Various unit forms

传统住宅："孤岛"，户与户之间交流甚少。几种有限的标准单元
联建住宅：社区活动频繁。多种单元形式

Fig.3.18订制性与个性 Customization & Identity

3.2 订制·个性

发展联建住宅项目背后的动机之一是由于目前市场上缺乏具有可替代性的住房解决策略，为了满足需求，部分人不得不变得主动，要为自己创造可能性，按照自身意愿来创造居住环境。因而联建住宅项目一个重要特色就是客户订制性，项目天然地满足客户的需求。这种订制性主要表现在两个方面，一是个性化的居住模式，一是个性化的居住户型（Fig.3.18）。

3.2.1 订制生活：个性化的居住模式

联建住宅项目旨在为用户量体裁衣，满足其个体需求，并能根据时间迁移做出相应改变。项目通常具备灵活的平面并满足无障碍建筑的相关标准，能够帮助实现多代居等灵活的居住模式。通过专门的解决策略，那些本来需要搬出城市以实现他们特定生活要求的使用者，就可以根据他们的需求，在内城中创造自己满意的住所。联建住宅项目最激动人心的部分正在于此，通过订制实现另一种居住模式，而这是普通住宅项目所不能实现的。不管是集体宿舍式住宅、老年住宅还是多代居住宅，作为普通类型住宅的补充，联建住宅项目极大地丰富了住宅类型和生活模式。

3.2 Customization & Identity

The driving force behind Co-housing is mainly rooted in the situation that the current housing market cannot offer an alternative solutions to satisfying people's changing living demands. In order to change the situation, people have to create the possibility by themselves to make their ideal living space. Customization is Co-housing's major characteristic and it presents mainly in two aspects: individualized living style and individualized housing unit layout (Fig.3.18).

3.2.1 Life Customization: Individualized living style

Many Co-housing projects aim at tailoring the users to living space and they can often adapt flexibly to the change of the users. Buildings usually have flexible plans and barrier-free designs which are suitable for multigeneration dwelling and other living models. By specialized strategies and customization, Co-housing group can practice an alternative living style in the inner city that the normal housing projects can hardly achieve. Collective dormitory pattern, retire house patter, intergeneration house pattern and other forms of housing living patterns, together form the abundant types of individualized living styles of Co-housing. As a supplement to the normal housing, Co-housing enriches types of housing and people's dwelling models.

06\ AL WiG 中老年社区养老生活 An Alternative Solution to Aged Support

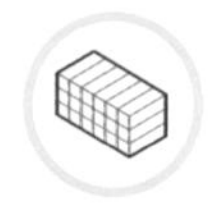

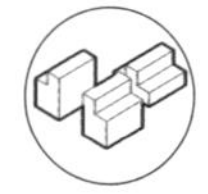

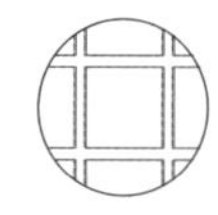

项目概况 Project Profile

项目名称 Project Name：	Allein Wohnen in Gemeinschaft (AL WiG)
项目地址 Project Address：	Falkstrasse 25
建筑设计师 Architect：	N/A
基地面积 Site Area：	N/A
建筑面积 Total Area：	550 m^2
项目成立时间 Founding Time：	2005
项目状态 Project Status：	完工 finished

Fig.3.19项目区位及实景 Project location & scene

Allein Wohnen in Gemeinschaft 项目位于柏林新科恩 (Fig.3.19)，该项目属于租住类型，其成员都是超过 60 岁的单身男女或是已经分居的夫妻。不同于普通的单身公寓的氛围，整个项目的口号是“即便你回家了也不会孤单！”，他们会定期集会并组织一些联谊活动。老人们居住在独立的公寓之中，通过共享公共活动室以及诸如洗衣机、汽车、报纸等生活资源，减少了对于环境资源的消耗，同时老人之间相互照顾扶持，效果要优于普通的养老院。

普通的养老机构由于花费太高，对于老人来说不具有可持续性与稳定性，而该项目则是对养老模式的另一种探索，能够为老人们永久性地提供价格适合且稳定的住所，同时也减轻了家属们的负担。而老年人居住习惯的长久性与稳定性也为房主提供了长期稳定的租金收入。

Project Allein Wohnen in Gemeinschaft (AL WiG) is located in Neukoelln (Fig.3.19) and is a rental type of Co-housing. All the group members are the single or divorced old man or woman over 60. Unlike the normal apartment for the single, the slogan of the project is “ you won't feel lonely even you are alone at home!”. They will organize theme meetings regularly to communicate with each other and learn something from each other. The members live in the independent unit and share the communal playroom and washmachines, cars, newspapers and other living materials. Members support each other and get additional helps from the volunteers. People feel better here than in the normal retirement house.

The normal nursing institution for the aged usually is not affordable and sustainable for the aged. This project develops an alternative ways of nursing the aged by providing them long-term affordable housing thus reducing the burden of the family. In return their stable living habitat also insures that the landlord can get long-term stable rent incomes.

07 Möckernkiez EG 多代居社区 Intergeneration Community

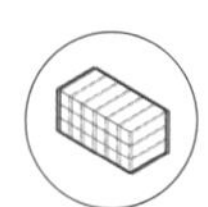
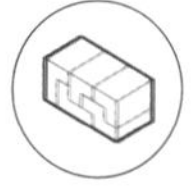
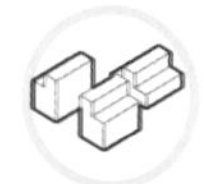

项目概况 Project Profile

项目名称 Project Name：	Möckernkiez EG
项目地址 Project Address：	Möckernstrasse/Yorckstrasse 24
建筑设计师 Architect：	DREES+SOMMER, BE BERLIN, R. DISCH SOLARARCH. ROEDIG. SCHOP ARCH. SCHULTE-FROHLINDE ARCH. BAUFROESCHE ARCH
基地面积 Site Area：	30000 m^2
建筑面积 Total Area：	62385 m^2
项目成立时间 Founding Time：	2007
项目状态 Project Status：	在建 under construction

Fig.3.20项目区位及实景 Project location & scene

如果说 Allein Wohnen in Gemeinschaft 项目是探索中老年通过互帮互助共同养老的模式，那么 Möckernkiez 项目则是在探索另一种养老模式——多代居社区，项目旨在通过家庭的力量实现养老，建成满足各代际人群居住需求的综合性社区。

Möckernkiez 项目是由柏林市民发起的，整个项目位于新轨道三角洲公园的东南角（Fig.3.20）。项目概念始于 2007 年 5 月克劳茨伯格地区社会民主党的一次内部会议，而到 2010 年 9 月才正式获得了 $3hm^2$ 的建设用地，计划到 2015 年底建成一个拥有约 460 户居民，集公共空间与商业空间于一体的，可持续发展的，可达性优良的，多文化混合的，代际交流密切的示范性社区。

项目的城市设计概念是在 Baufrösche architects and urban planners GmbH 和 BE Berlin GmbH 两家设计公司的配合下，由即将入住的居民共同参与完成的。该地块比周边高出 4m，这也成为该用地的一大特色。为了充分利用这一特色，社区由不同类型房屋所组成的小尺度集群构成，以此创造一个拥有较高居住品质的周边式围合小区。与 Yorckstrasse 大街相平行的是一长条建筑，并且向公园方向长出了三个分支。与 Möckernstrasse 相平行的体块则主

If project Allein Wohnen in Gemeinschaft is similar with the retirement house pattern, then project Möckernkiez tries to solve the issue by intergeneration community. This project aims to create a comprehensive modern community for all generations and to support the aged with the power of the families.

The project is located on the edge of the new Gleisdreieck-Park in Kreuzberg (Fig.3.20) and is initiated by Berlin citizens. With 3 ha areas of land, the project plans to build an ecologically sustainable, accessible, cross-culturally and socially inclusive living community with about 460 apartments, common areas and commercial areas by the end of 2015.

The urban planning concept was developed in a participatory process with the involvement of future residents by the Baufrösche architects and urban planners GmbH and BE Berlin GmbH.

Compared with the surrounding areas, Möckernkiez like the entire area of Gleisdreieck-Park is increased by four meters. For this characteristic the community was developed as

Fig.3.21 项目总平面图 Masterplan

要用来阻挡基地东侧对于无车居住区的干扰。面朝公园的则是几幢拥有私人花园的行列式住宅，这些建筑主要是东西朝向，而东北侧的空地则由一幢点式住宅所填补。尽管是周边围合式，住宅区仍然对外开放。它提供了 6 个出入口供人穿越，并提供了直达轨道三角洲公园的快速通道。主入口位于 yorckstrasse 靠近约克桥的位置，设置了标志性的大台阶以及坡道系统（Fig.3.21）。

本项目也是柏林联建住宅项目中最大型的社区项目之一，从户型到社区再到邻里，可以说都是高度订制的，庞大的联建团体也使得项目本身的进程十分缓慢，是对联建住宅项目适宜规模的又一次先锋性探索。

small-scale clusters with different house types thus creating a charming sequence of rooms and outdoor spaces with high living quality.The community forms a perimeter block next to Yorckstrasse and Möckernstrasse. The elongated block of buildings with three " wings " orienting to the park parallels to the historic escapeway Yorckstrasse. Another block, which parallel to Möckernstrasse protects the car-free residential area from the east disturbance. Facing to the park are row houses with private gardens, and the northeast site is filled by an independent free-standing housing. The buildings are predominantly east-west oriented. The community offers six entrances that invite people to cross and offer quick access to the Gleisdreieck-Park.The main entrance is at Yorckstrasse close to the York bridge in the form of a representative outdoor stairs with a ramp system. (Fig.3.21)

This project is one of the biggest Co-housing projects. It is highly customized from the housing unit layout to the community layout. Such a “huge” Co-housing group (nearly a thousand) makes the projects develop quite slowly. It is another pioneer practice of the Co-housing in terms of its scale.

08\Malmoeer strasse29

纯粹的集体生活
Pure Collective Living Style

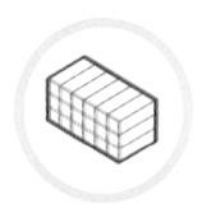

项目概况 Project Profile

项目名称 Project Name： M29
项目地址 Project Address： Malmoeer strasse 29
建筑设计师 Architect： CLEMENS KRUG ARCHITEKTEN
基地面积 Site Area： 800 m^2
建筑面积 Total Area： 1012 m^2
项目成立时间 Founding Time： 2008
项目状态 Project Status： 完工 finished

Fig.3.22项目区位及实景 Project location & scene

柏林有很多以租赁模式为基础的联建住宅项目，但M29项目的特殊之处在于其成员并没有参与住宅的建设规划阶段而是直接入住的。该项目共居住了20人，这其中包含了3个小孩。整个项目的核心就是定期举行的全体会议，主要是讨论一些公共事宜，所有的决议必须达成共识以确保所有人都满意。他们想要平等公平地、对彼此负责地共同生活。建筑是由底层的独立卧室、分布每层的小厨房（用于吸烟或者荤食）以及位于顶层的一些公共空间如公共厨房、图书馆、工作坊和酒吧等组成。

项目各种类型的贷款都以房租的形式分期缴付，每个月每个人需缴纳300欧，此外每人还需缴纳100欧（小孩50欧）的费用作为公款，用来支付日常开销，同时他们也会定期举办各种活动来赚一些零用，如定期在每月第一个周四晚上开放的公共食堂。

联建住宅项目的另一个特点就是大多数项目都不是孤立的，各项目通过一些组织相互联系、互相扶持与帮助。M29住宅项目就是著名的Mietshaeuser Syndikat组织的成员，该组织成立于弗莱堡，致力于将住房永久性地隔离出市场并防止投机倒卖。所有建筑都不是私有的而是自助式管理、由责任有限公司所有，持股者包括居民以及该组织的成

In this project, 20 people live in common. In the group there are already three children and they expect that there will be more. The center of all processes is the regular plenary session, in which all discuss the common affairs together. All decisions are taken by consensus to ensure that all are well satisfied with the result. They want to live together with equal rights, take responsibility for one another. They see themselves as organized on a democratic basis, anti-sexist, anti-racist and anti-fascist project.

The house is composed of the individual bedrooms (underground) as private spaces, several smaller kitchens for special needs (such as smoking or meat meal), communal kitchen, shared library & workshop and bars as shared spaces upstairs.

As this project is in the membership of Mietshaeuser Syndikat, the basic objective ensured is that the house is permanently deprived of the market and possible speculation. Many small and large direct loans funded the construction and these loans are amortized over the lease. Monthly they should pay 300€ each person to cover these

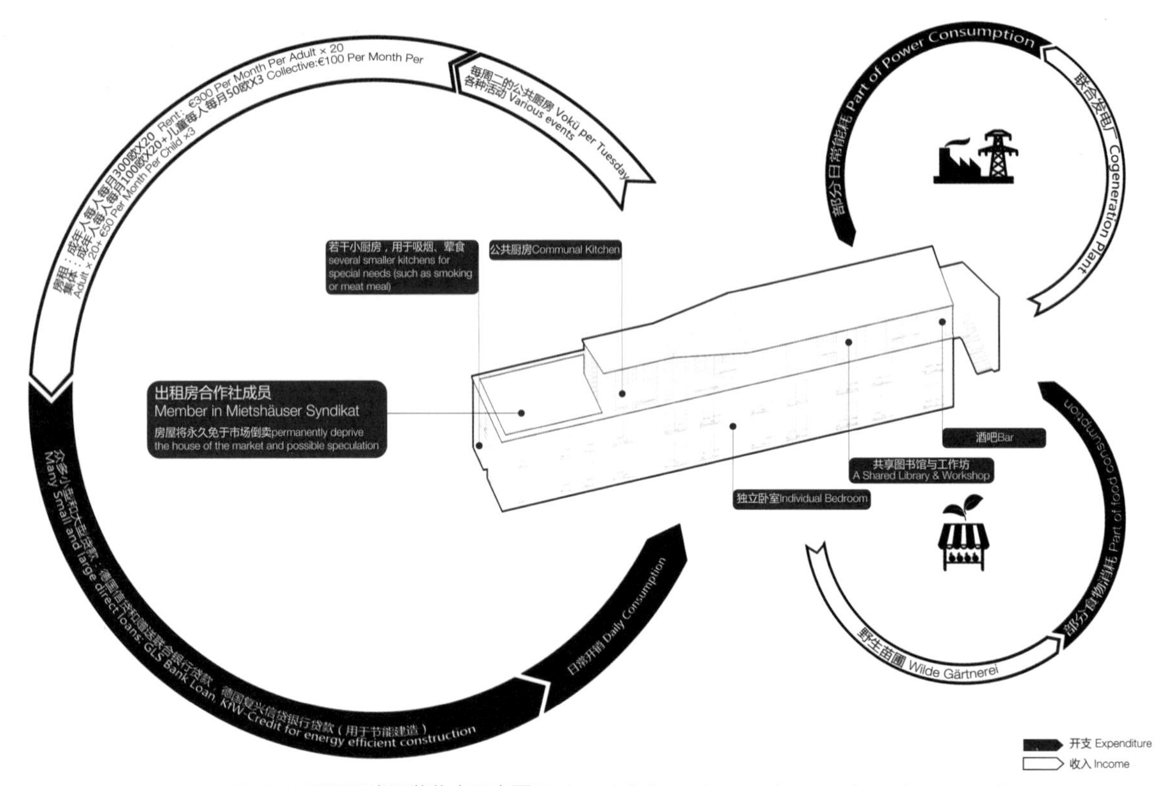

Fig.3.23 项目日常运营收支示意图 Project daily incomings and outcomings demonstration

员。所需资金都是靠向亲戚和朋友借钱。项目的花费不是按比例承担的，这意味着有能力多付的就多付，其他人就少付，从而帮助那些经济能力较差的人加入。其他贷款的花费作为房租均摊到每人头上。该项目的基础租金是头三年每平米 5.9 欧，之后为每平米 7.2 欧，其中一部分租金交给出租房合作组织用以资助其他新的项目。人们可以通过支付一次性会费（250 欧）加入该组织，然后参加它的新项目。房租长期保持不变，未来获得的任何利润都将用于资助新项目。该组织成员之间互相分享项目运行经验，并且能够在经济上援助一些刚刚兴起的项目。此外，M29 项目还参与了其他一些项目，为其他联建项目提供宣传平台等（Fig.3.23）。

loans. Besides each person should pay 100€ and 50€for each child to cover the daily consumption.They will also do various events to make themselves known like the VoKue and earn some little money at the same time.

Another feature of Co-housing projects is that they are in the extensive network. They are also part of other cycling projects. M29 also works with local co-generation plant and the Wilde Gaertnerei [1] to get cheaper living resources（Fig.3.23）

1 Wilde Gaertnerei is an independently founded, since 2012 community supported organic pioneer farm thirty kilometers from Berlin. Source: Wildegartnerei.blogspot.de

3.2.2 订制空间：个性化的居住户型

联建住宅项目归根结底是住宅，是满足团体成员各种居住需求的住宅，不管它倡导的是哪一种生活方式，最终都需要以空间的形式落实，所谓众口难调，势必在同一个项目中存在着多种户型或者一种可变性极大的灵活户型。

3.2.2 Space Customization: Individualized Housing Unit Layout

The nature of Co-housing is housing that can satisfy all the living demands of its group members. No matter what kind of living style it advocates, it must base on the building itself and the housing units. Usually in one project it must offer many unit-layouts or several flexible units that can change to meet different users' demands.

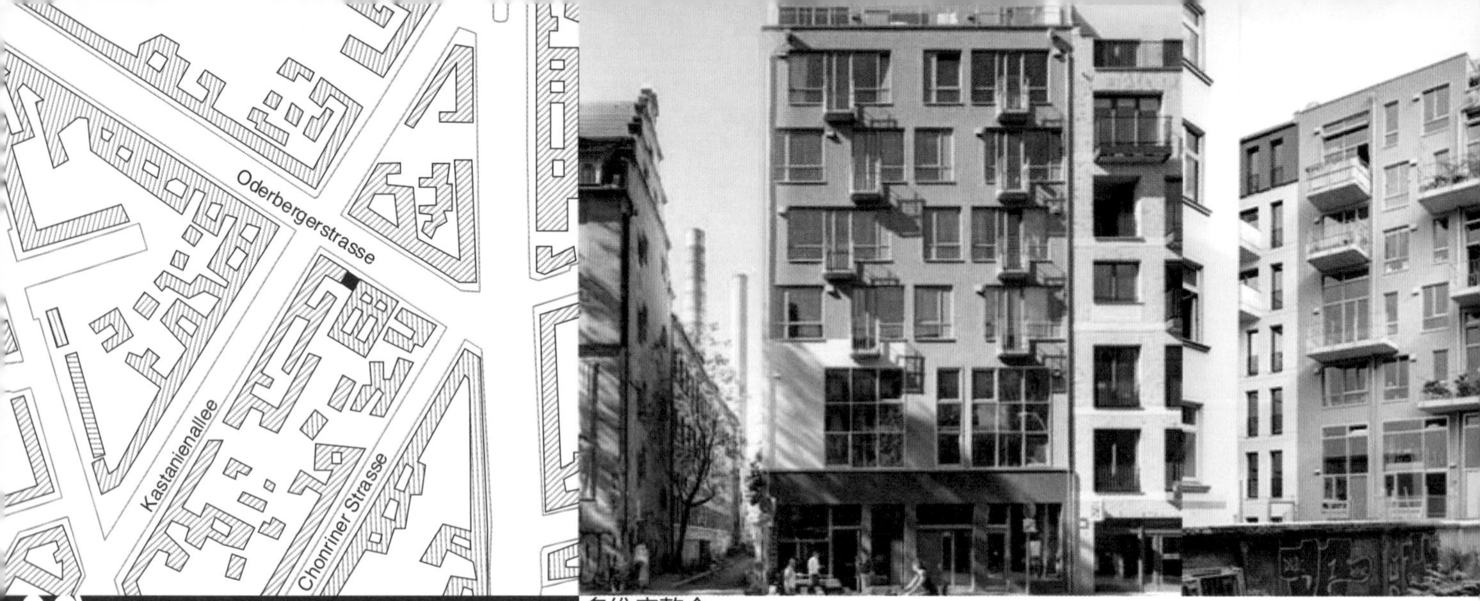

09 Oderberger Strasse 56

多维度整合 Integration of Multiple Dimensions

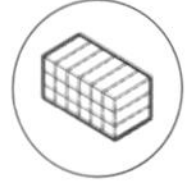
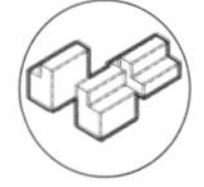

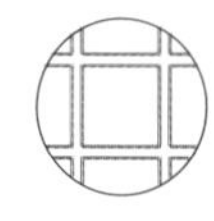

项目概况 Project Profile

项目名称 Project Name：Baugruppe Oderberger Strasse 56
项目地址 Project Address：Oderberger Strasse 56
建筑设计师 Architect：BAR ARCHITEKTEN
基地面积 Site Area：315 m^2
建筑面积 Total Area：1093 m^2
项目成立时间 Founding Time：2003
项目状态 Project Status：完工 finished

Fig.3.24项目区位及实景 Project location & scene

本项目位于 Oderberger Strasse 56 号 (Fig.3.24)，这里曾是一块荒地，如今已经建成了一栋拥有五套公寓，五套工作室且功能可变的高密度实验住宅楼。

整个建筑七层半高，面向街道有较小的出入口，底层为商业，紧邻柏林市政浴场。仅仅通过横跨二三层的工作室的玻璃窗就可以看出本栋建筑不仅仅是住宅，更是一栋高密度功能嵌套型综合体。五套工作室之一为建筑事务所，在建筑立面上通过浅色墙漆来体现其空间体量。最大的工作室面积为 45m^2，跨越两层，虽然面积小，但仍然配有厨房和卫生间，隔壁房型则更小（分别为 31m^2，33m^2 和 35m^2）。

小户型是该项目对于柏林内城住房市场强有力的反击。在房地产投资商看来，住房似乎应该“更大一点”，越多的平米数意味着越多的金钱。在本案中，更多的精力和财力都花在了创造一系列特殊结构上，例如不同的层高（2.10m~4.20m）、一套公寓拥有四个楼层、套内楼梯、空中走廊及部分订制的家具等等。最多的是跃层带阳台的户型，朝南面向庭院，主要用作起居空间。户型之间相互咬合，在高度上错半层，进深上也错半。最大户型面积为 125m^2，其空间紧凑配有四个房间，同时也考虑了不同住户家庭构成，这些房间可以很容易地被分隔，例如供给祖父

The project is located in Oderberger Strasse 56 (Fig.3.24). Here used to be no-man's land and now stands a high-dense function-changeable experiment housing with 5 apartments and 5 studios. Due to the land limit, the architects co-work with the Co-housing group to calculate the volumes and finally achieve this volume-interlaced building with high space utilization. After 3 years the whole project finished in 2006.

The Building has seven and a half storeys and there is a small exit towards the street.The underground floor is for commercial use and next to the municipal bath-hall. One of the five studio is an architecture design office and the corresponding volume's facade is coated with light paint to distinguish itself from the other parts. The largest studio has an area of 45m^2 and it is duplex. And the others are 31m^2,33m^2 and 35m^2.

The housing units of small area are designed to against the housing market in inner city. In the view of the real estate developers, the housing units should be larger and larger and then they can sell them at higher and higher price and

m2
客房 guest room 18
套间 room 5a 78
套间 room 5b 37
115
套间 room 4a 83
套间 room 4b 45
128
套间 room 3a 81
套间 room 3b 27
108
套间 room 2 76
套间 room 1a 82
套间 room 1b 43
125
工作室 studio A 45
工作室 studio B 35
工作室 studio 1 31
工作室 studio 2 33
工作室 studio 3 33
咖啡厅 cafe 1 51
实验空间 experiment space 5
商店 shop 43
音乐室 music room 28

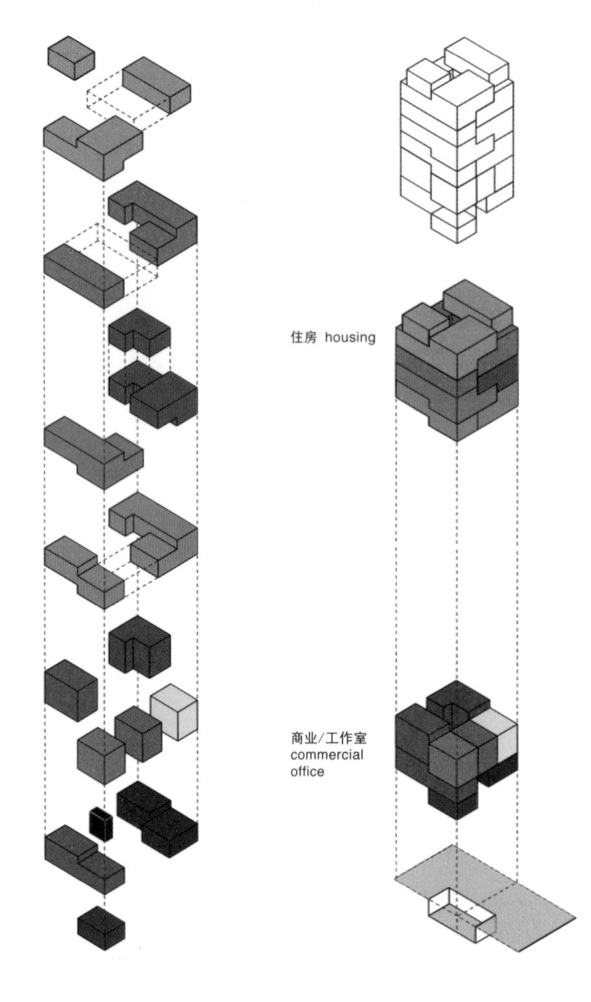

首层|电梯间前是一小型画廊；音乐室前为采光庭院。
Groundfloor I
the small room before the lift is a gallery; there is a light-patio in front of the musicroom.

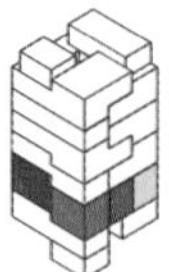

一至二层|五间工作室：面向庭院的三间可由预制甲板和钢梯到达。
1-2 floor I five studio: three studio facing the courtyard is accessible by the pre-set deck with steel stairs.

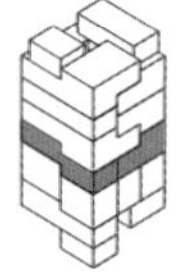

三层|125m²，其中40m²可以被分隔出去。木制楼梯连接不同标高。
3 floor I 125m², of which 40m² can be seperated.Wooden stairs connect different levels.

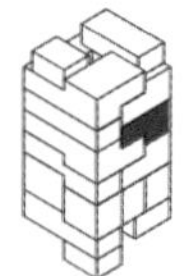

四至五层|110m²：厨房之上挑出的画廊联系了上下的生活区域。
4-5 floor I 110m²: The gallery, which is on the top of kitchen, connects the living area up and down.

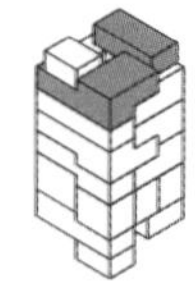

七层|从画廊延伸过来的悬挑狭长甲板将空间垂直划分为图书室和厨房，各自拥有2. 1m层高。
7 floor I The suspended narrow deck divides the space into library and kitchen vertically with 2.1m height respectively.

Fig.3.25 空间组织关系示意图 Space Organization demonstration

母、护理人或是租客。再分隔后的房间可以从半平台位置预留的户门进入（Fig.3.25）。

本案空间种类丰富，除去公寓和工作室外，还配有带客房的共享屋顶露台、“实验空间”（朝向街道的非营利性画廊）以及音乐室。

值得一提的是本案的组织模式，既非传统的联建模式也非开发商模式。主要有以下特点：

（1）进程缓慢以及预筹资模式：早在2003年用地就已购买完毕，买主包括了建筑师；

（2）所有制混合（Fig.3.26）：房屋建设费用一半由3套公寓的业主支付，另一半由租户以预支租金的形式支付包括工作室以及剩下的一套公寓租户；

（3）建筑师角色的延伸：建筑师作为租户成员还承担了门卫以及建筑管理的职责。

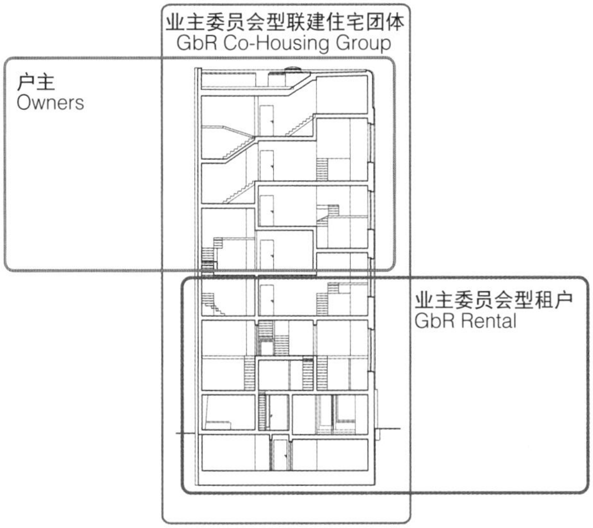

业主委员会型联建住宅团体
GbR Co-Housing Group
户主
Owners
业主委员会型租户
GbR Rental

Fig.3.26 项目所有制模式示意图 Ownership model demonstration

don't consider about their flexibility. In this case more efforts are put on creating a series of features on the structure: different story height (2.1m-4.2m), an apartment with at most four stories, sky corridors, customized furniture and so on. Units interlaced with each other and have half-story height difference and half depth difference with each other. The largest housing unit has an area of 125m^{2}with 4 rooms. It also considers about the different family composition, so the rooms can be subdivided into independent apartments with exclusive entrances (Fig.3.25). Besides of the apartments and studios, the groups share the roof terrace with guest rooms, the experiment space (mini-gallery) in the lobby area and the music room in the underground floor.

What is worth of pointing out is its Organization forms, neither the traditional Co-housing model nor real estate model. It can be summarized as follows:

(1) Slow Process and Pre-financing: the site was purchased in 2003 and the buyers include the architects;

(2) Mixed Ownership (Fig.3.26): half of the construction expenditure is paid by the 3 apartment-users and the other half is paid as the rent by the 5 studio-users and one apartment-tenants;

(3) Extension of the architects' role: the architects act not only as the tenants but also the gate keeper in charge of the building management.

02 \ Zwillinghäuser

标准化中的个性化
Individulization in Standardiszation

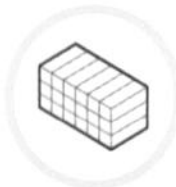
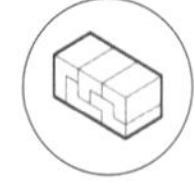
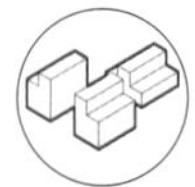

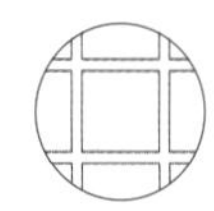

项目概况 Project Profile

项目名称 Project Name： Zwillinghäuser
项目地址 Project Address： Schönholzer Str.10A/ Ruppiner Str. 43
建筑设计师 Architect： ZANDERROTH ARCHITEKTEN
基地面积 Site Area： 1466 m^2
建筑面积 Total Area： 4546 m^2
项目成立时间 Founding Time： 2005
项目状态 Project Status： 完工 finished

Fig.3.27项目区位及实景 Project location & scene

在如何体现居住者个性方面，Zwillinghäuser 项目提供了不同于 Oderberger Strasse 56 项目的另一种方式，即先研究灵活性较高的标准平面，然后根据住户的实际需求进行变动。

这两栋七层高的双子住宅楼平面呈镜像关系，每层平面都有三面景观朝向。楼层稍低的户型，由于南面有可能被内庭院里的建筑所遮挡，起居空间被置于靠近街道一侧，而在稍高的楼层起居空间则仍然位于朝南面。建筑的内部结构通过立面的设计就能反映出来。

本案是多代居联建住宅项目，12 户套型除了满足正常的三口之家的生活需求外，还分别为年轻的六口之家，单身女性以及老年夫妇提供了适宜的居住空间。首两层作为跃层户型配备私人花园，最上层设计成了经典的阁楼同时提供了屋顶露台。标准层面积约为 117m^2，被分隔为起居空间，服务性空间（厨卫浴）以及户外空间。比较突出的细节是在朝向广场的一角采用了倒角形式，增强了整个社区的亲和力 (Fig.3.27)。

本案的基本结构为框筒结构，这使得它能够根据不同居住人数来调整房间的功能与布局，空间的使用与分配十分灵活。每一户都拥有位于阳面的起居室，通过打开折叠门可

In terms of how to realize customization of housing units, project Zwillinghäuser offers another strategy: firstly design a highly-adaptative standardized unit and then under fixed rules the residents can transform it according to their demands. This transformation can happen within the unit or can be achieved by putting several standardized units together as a “bigger one”.

These 7-storey twin housings have the mirrored plans and each floor has three sides of view. On the lower floors the living areas are arranged in the street side because the south face may be shaded by the buildings in the inner courtyard. On the higher floors the living areas are arranged in the south face.

This case is a multigeneration housing with 12 units, which can not only meet the normal nuclear family’s demand but can also adapt to the young family of 6 people, single woman and elderly couple. On the first two floors there are duplex apartments with private gardens while on the top floor there are classical loft apartments with roof terraces. All the units are based on the standardized unit of 117m^2 which

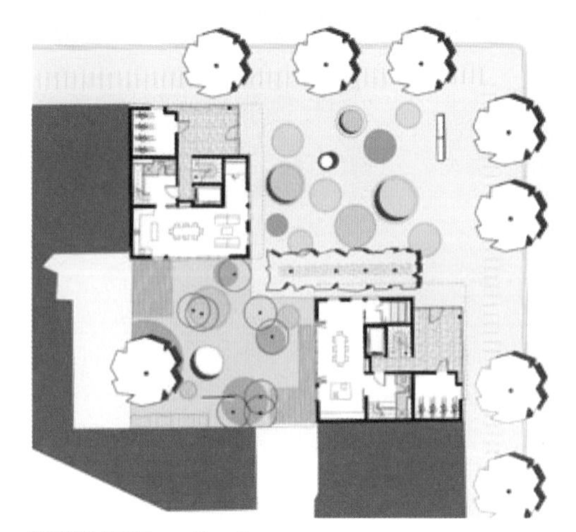

首层组合平面 Groundfloorplan

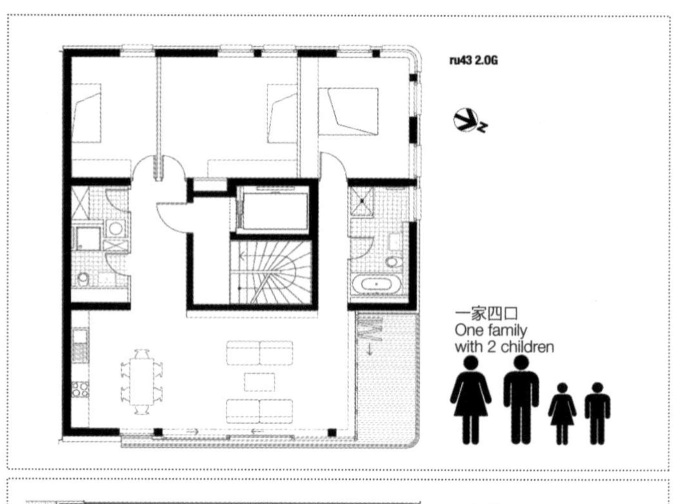

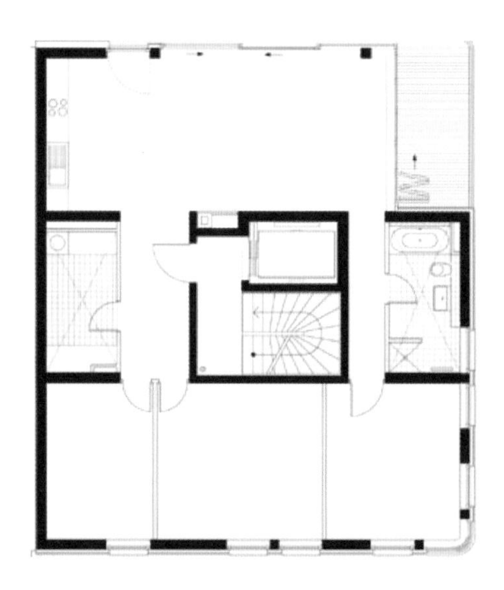

标准层平面 Standard Floorplan

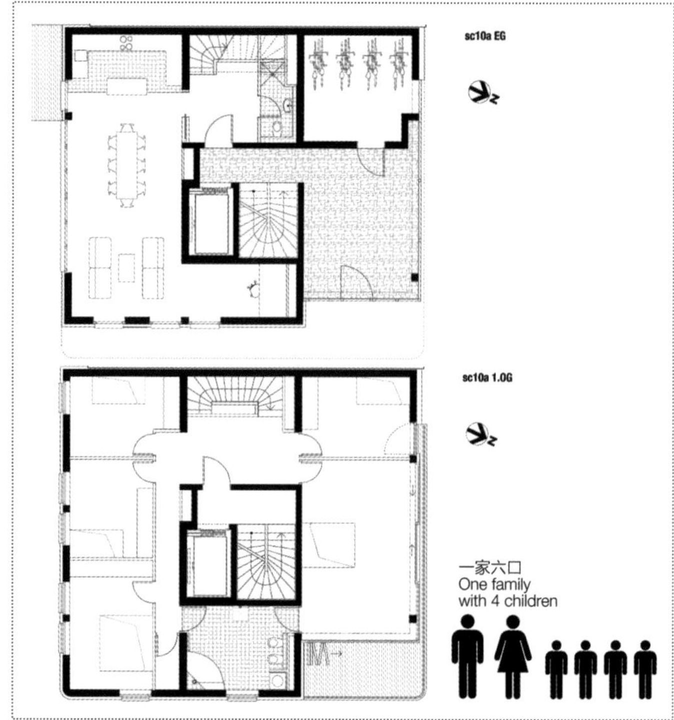

Fig.3.28灵活的平面 Flexible plan

与阳台直接连通。在一些户型中，使用者可以围绕交通核环通四周，而其他户型大部分住户则选择将室内分成两翼，一翼归父母使用，一翼归孩子使用，最大程度保证了各自的私密性（Fig.3.28）。

consists of living area, service area (kitchen, bathroom and toilet) and outdoor space(Fig.3.27).

This case adopts frame-tube structure which can achieve flexible plan adapting to different users. The unit type ranges from two rooms to four rooms, from unit with independent kitchen to unit with open kitchen. Every unit has the living room facing the south. By opening the folding door the living room can extend to the balcony. In some units the users can go around the circulation hub while others are divided into two “wings”: one for the parent and the other for the children, ensuring the respective privacy.(Fig.3.28)

04 R50 基于模块构件的设计游戏 A Design-game based on Modular Components

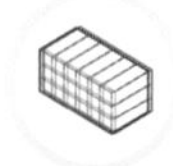
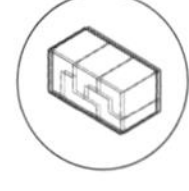
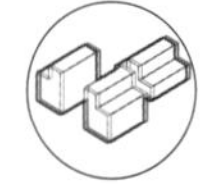

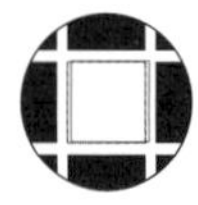
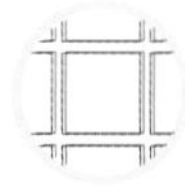

项目概况 Project Profile

项目名称 Project Name： R50
项目地址 Project Address： Ritterstrasse 50
建筑设计师 Architect： IFAU+JESKO FEZER HEIDE & VON BECKERATH ARCH
基地面积 Site Area： 2055 m^2
建筑面积 Total Area： 2780 m^2
项目成立时间 Founding Time： 2010
项目状态 Project Status： 完工 finished

Fig.3.29项目区位及实景 Project location & scene

本案例建筑设计的方法值得进行讨论。具体的过程如下：建筑师与居民先进行沟通，居民需要画出自己理想的居住空间结构图，包括功能空间的大小，组成以及它们之间的关系，形成如 Fig.3.30 所示的居住空间评估参考系统。

根据这一反馈，建筑师开始制定“游戏规则”：首先确定整个建筑体量的结构形式及其基本框架，本案为框筒结构，由中心的服务核（包括交通核和开放式厨房）和周边的承重墙结构组成。然后设计一系列模数构件（诸如分隔墙、门窗框、门窗扇等），每一类型都有多个型号可供选择，而这些构件的选择交由居民决定，包括如何划分内部空间、开窗数、开窗大小等。这样就形成了丰富的立面以及满足住户个性化需求的居住空间 (Fig.3.31、Fig.3.32)。

The participative design process of R50 is worth of discussion, and it can be described as follows: firstly the architects communicated with the residents and the residents were told to draw out their desired room sizes and the space relations of the functional rooms (Fig.3.30). Based on this methodological evaluation of residential spatial reference systems, the architects started to design “game rules”. They defined the frame-tube structure system and the general frame of the building, which allows changeable plans. Then they developed a series of modular components including partition walls, windows & doors and so on and each type has several options. Based on these rules, the residents can choose the components according to their willing to divide the units including numbers and sizes of the rooms and windows, thus creating different individualized housing units and individualized building facade. (Fig.3.31, Fig.3.32)

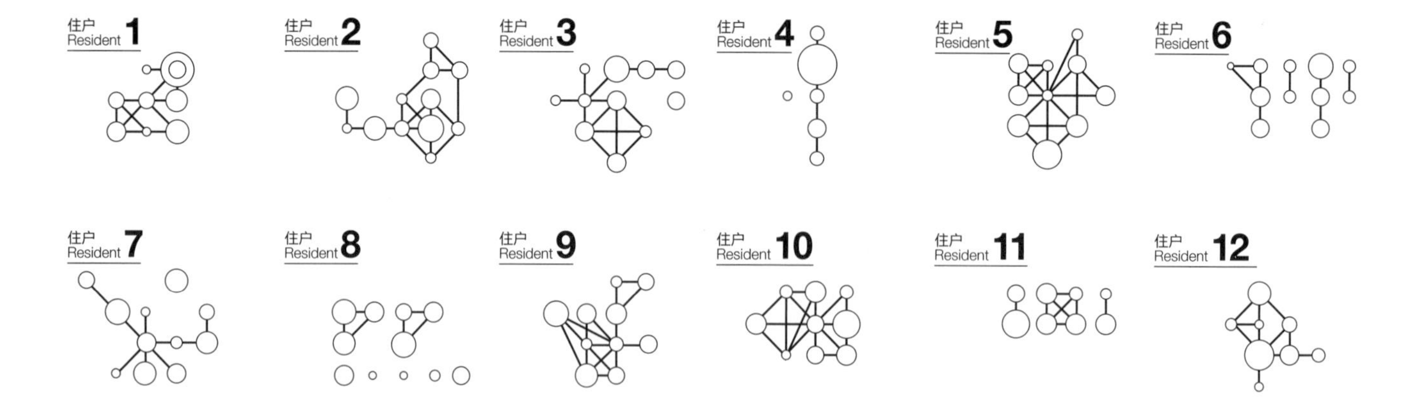

Fig.3.30居住空间评估参考系统：基于居民获得的理想空间的尺寸与空间关系
Methodological evaluation of residential spatial reference systems：desired room size and space relations based on residents

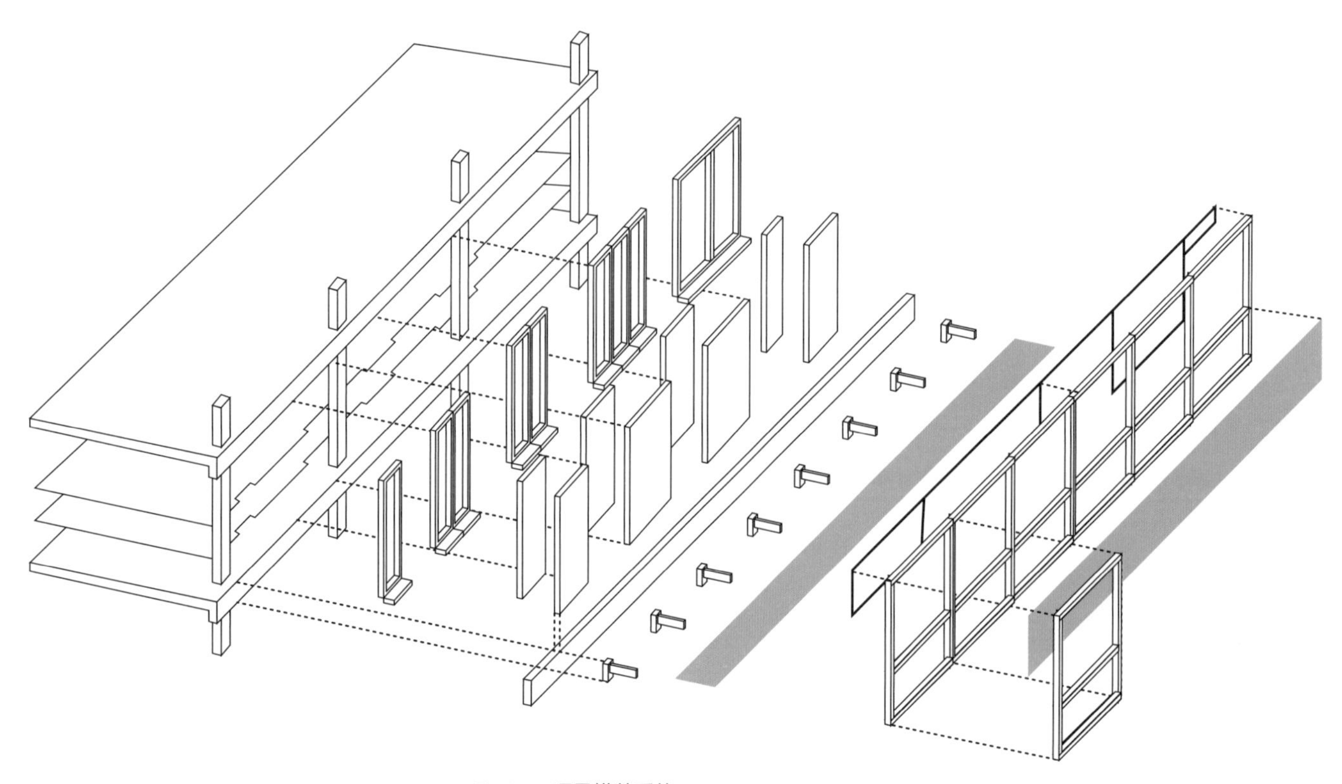

Fig.3.31 项目模数系统 Module system

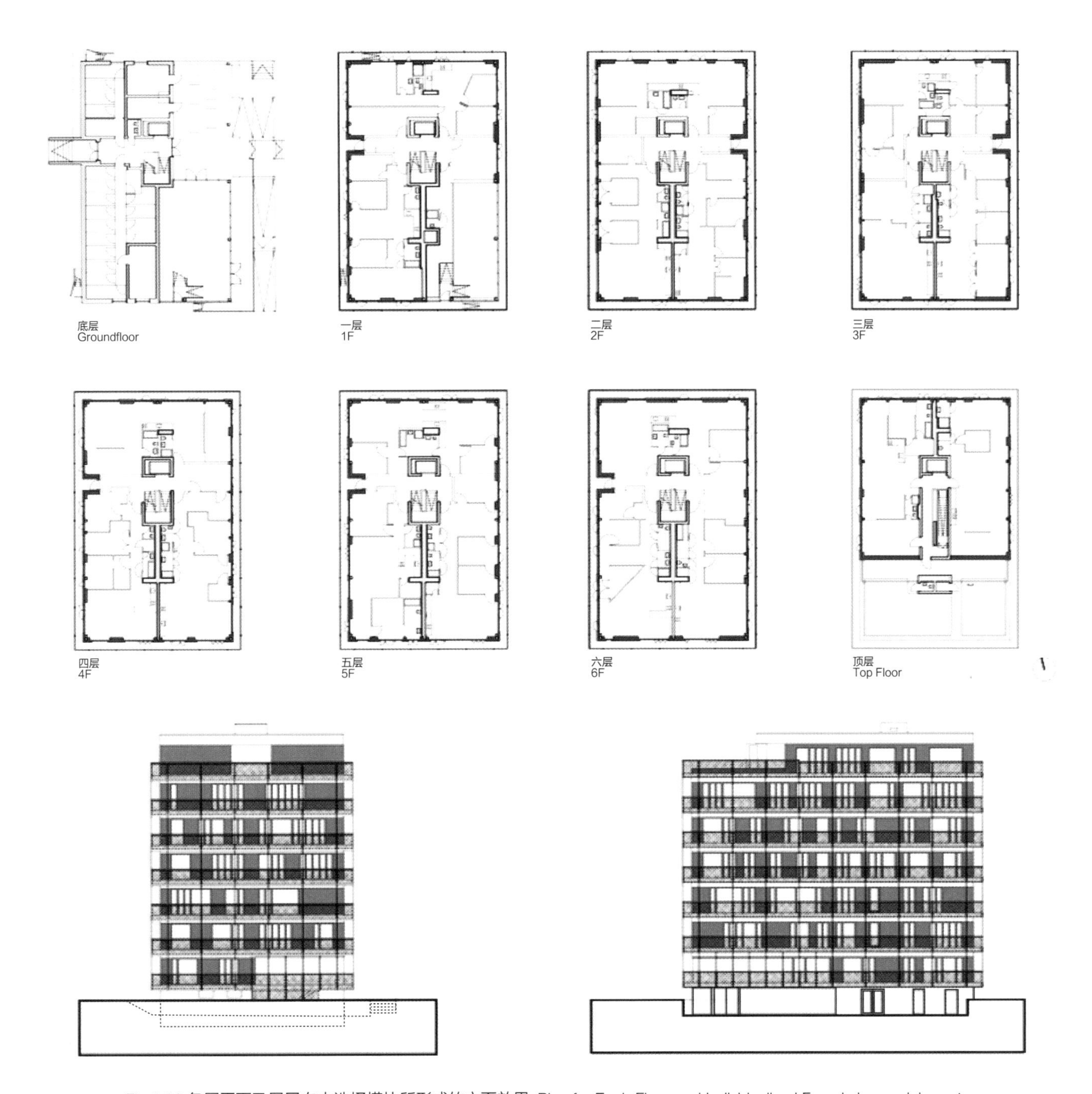

Fig.3.32 各层平面及居民自由选择模块所形成的立面效果 Plan for Each Floor and Individualized Facade by module system

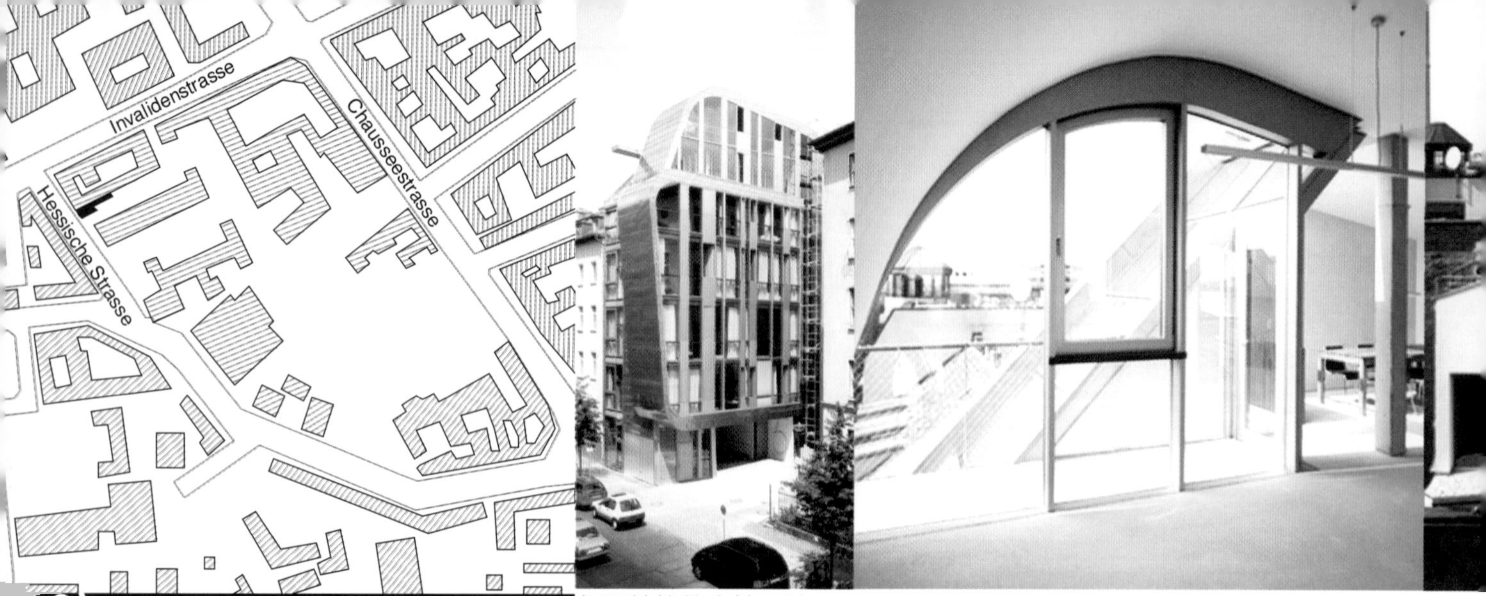

10\ Slender+Bender

极限地块的个性设计
Individualized Design on “extreme” sites

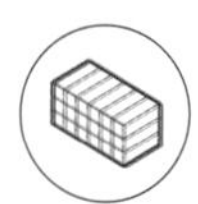
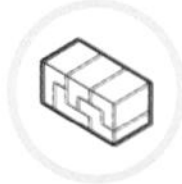
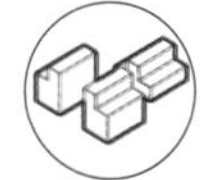

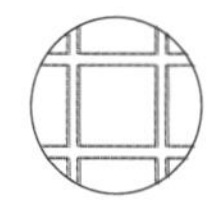

项目概况 Project Profile

项目名称 Project Name：	Slender+Bender
项目地址 Project Address：	Hessische Strasse 5
建筑设计师 Architect：	DEADLINE ARCHITEKTEN
基地面积 Site Area：	N/A
建筑面积 Total Area：	N/A
项目成立时间 Founding Time：	1999
项目状态 Project Status：	完工 finished

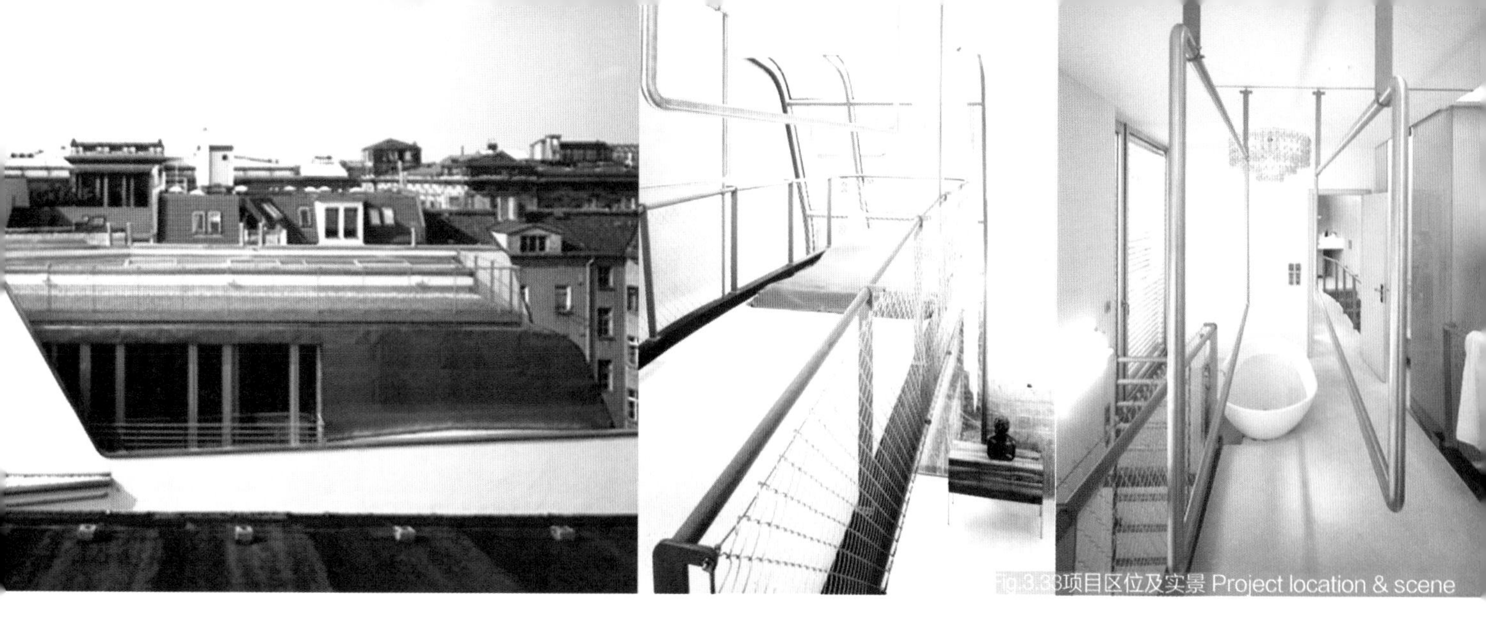

Fig.3.33 项目区位及实景 Project location & scene

Slender+Bender 项目基地十分狭窄 (Fig.3.33)，按照常理来说是无法使用的边角空间，但是建筑师却给出了创造性的解决策略：项目本来是针对一栋退界废弃的四层老建筑的改建项目，但是建筑师并没有采用扩建或是补建的方式来进行设计，而是创造了一种新的城市居住类型，在同一个场地里整合了多种建筑类型，背面紧贴那栋老建筑 (Fig.3.34,Fig.3.35)。

建筑总共 7 层，9m 宽，进深 14m，包括 14 户 $40m^2$~$45m^2$ 的 miniloft，一户 $130m^2$ 的家庭公寓，一间 $100m^2$ 的工作室、一间 $30m^2$ 的商店以及 $80m^2$ 的露台花园 (Fig.3.36) 。

这种项目是传统开发商不会涉及的，建筑师选择这种极限地块，设计出与之相应的居住空间，然后再来寻找与之相应的住户。在本案中，建筑师充当了开发商的角色，这也是联建住宅项目的另一种尝试。

Project Slender+Bender is the combination of new buildings and old buildings. They are located in an extreme narrow piece of land which is a wasteland in the normal sense(Fig.3.33), but the architects give an innovative design solution. This project is based on a 4-storey old building. Instead of simply transforming the old building, the architects create a new type of urban living and integrate many types of space forms close to the old building(Fig.3.34, Fig.3.35).

The 7-storey building is 9m wide and 14m deep containing 14 miniloft-apartments of 40-$45m^2$, a family apartment of $130m^2$, a studio of $100m^2$, a shop of $30m^2$ and a terrace garden of $80m^2$. (Fig.3.36)

This project is the type that the traditional developers wouldn't like to invest in. The architects designed the living space according to the extreme conditions of the site and then found the corresponding residents to form the Co-housing group, which was another experiment of Co-housing.

Fig.3.34 项目改造前实景 Scene before renewal

Fig.3.35 项目改造后实景 Scene after renewal

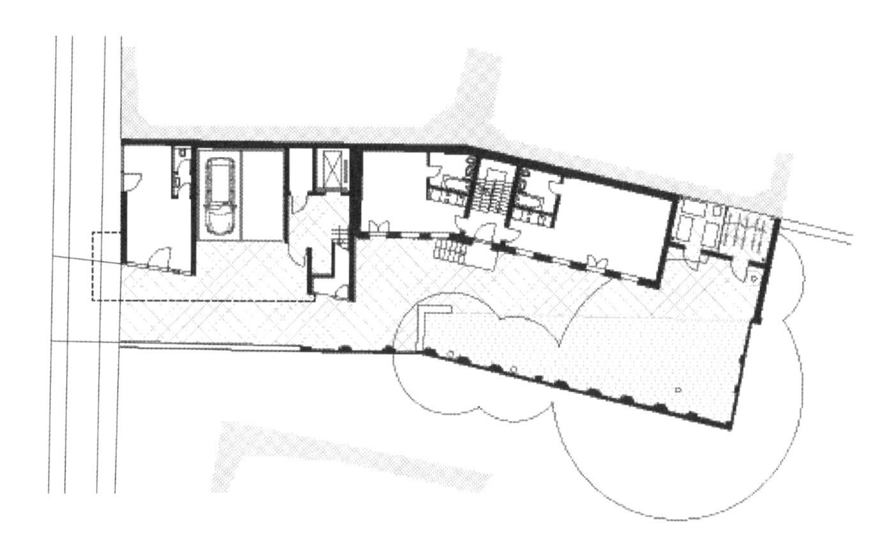

Fig.3.36 项目平面及剖面 Plan & Section

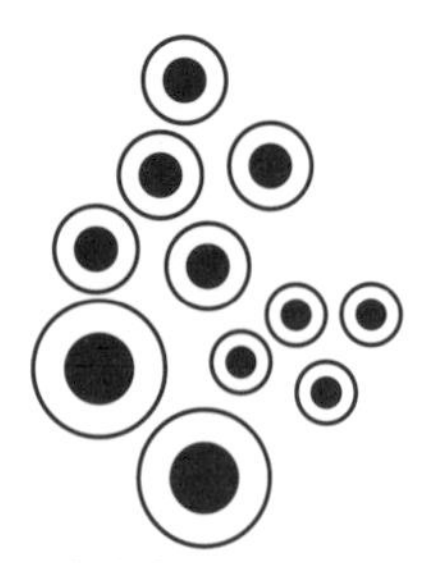

传统住宅
"孤岛"：户与户之间交流甚少

Traditional Housing
Isolated isles:little communication between neighbours

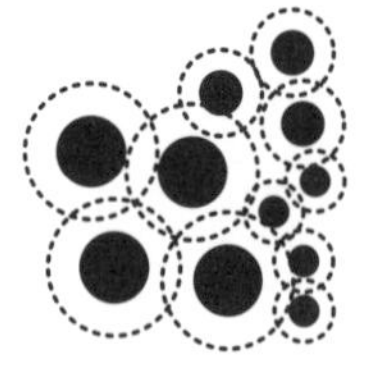

联建住宅
社区活动频繁

Co-Housing
Strong communication between neighbours

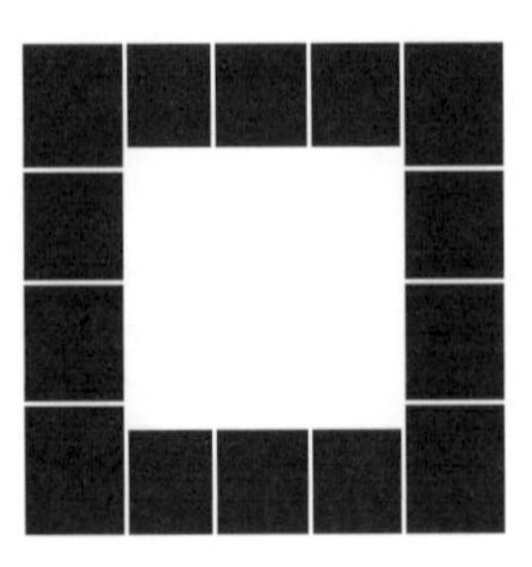

传统住宅
几种标准单元

Traditional Housing
Several standard units

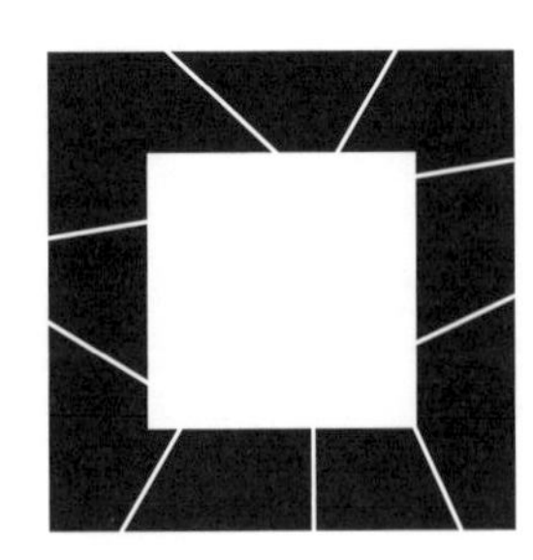

联建住宅
多种单元形式

Co-Housing
Various unit forms

Fig.3.37联建住宅与传统住宅在订制性与个性上的区别
Difference between Co-housing and traditional housing on Customization & Identity

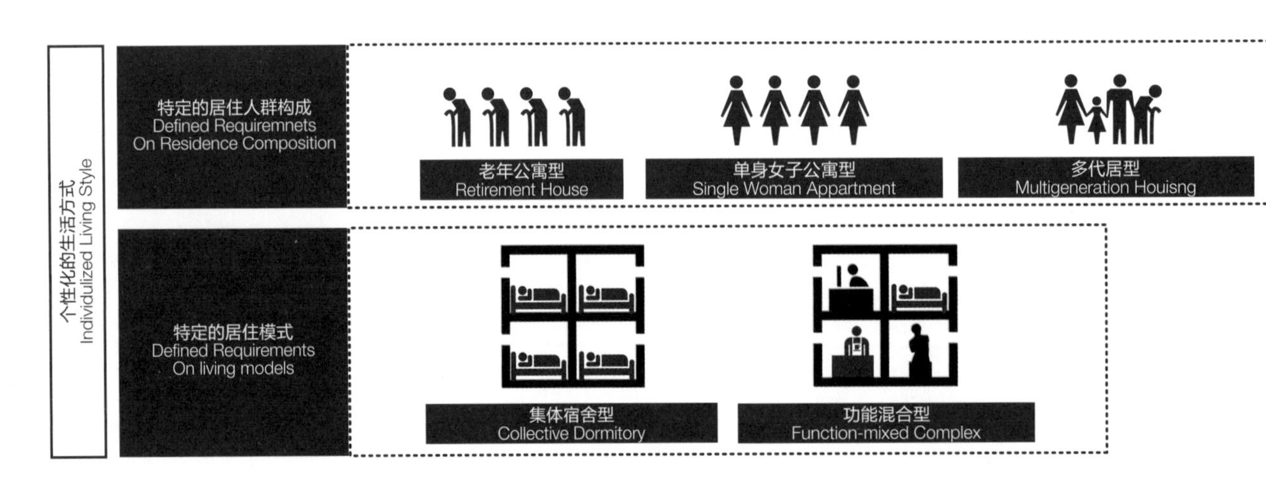

Fig.3.38个性化生活方式的探索 Exploration on individualized living style

3.2.3 小结

在传统房地产住宅项目，尤其是高层住宅中，居民都被禁锢在户内，如同生活在一个个“孤岛”一般，缺乏人际交流，这一方面是受现代生活模式的影响，大量的人群被集中到一个地方，个体淹没在人群之中，“来不及”与他人建立联系，另一方面正如扬•盖尔在“交往与空间”一书中所说：“在许多情况下都可以发现，物质环境能不同程度地影响居民的社会状态。物质环境自身可以设计成阻碍乃至扼杀所要求的接触形式，从建筑着手完全能够做到这一点”[1]。虽说开发商并不想扼杀居民的社会交往，我们依然能够看到社区当中大量公共空间的存在，可大部分情况下这些空间并没有被有效地利用。而另一方面，传统的住宅项目由于成本考虑，往往只有几种标准平面，通过反复组合来形成变化，而在联建住宅项目中，甚至每户的平面都不相同，而其实现的手段多样并不以高额的成本为代价（Fig.3.37）。本小结从居住模式与居住空间两个层面入手，总结联建住宅项目是如何通过订制实现个性化的生活从而改变传统住宅所面临的尴尬局面。

联建住宅项目追求个性化居住模式的探索可以归结为以下两类（Fig.3.38）：

（1）**对居住人群构成有特定要求**：这类项目往往会确定入住人员的构成，这可以发生在结成联建团体的阶段，也可以发生在项目完成后对新进成员的筛选阶段。主要的探索形式有老年公寓型、单身女子公寓型、多代居型等，体现了对于社会上某些弱势人群问题的关注。

（2）**对生活方式有特定要求**：这类项目对于入住人员的构成无特别要求，主要是希望探索某些另类的生活方式。主要的探索形式有集体宿舍型、功能混合型（居住、办公、文化、商业的混合）。

在联建住宅项目中由于多方都要主张自己的喜好与需求，而本着公平平等的原则，最后所形成的设计方案是在多方利益角逐情况下完成的，且各方对于结果都是满意的。为

1 扬•盖尔著，何人可译，交往与空间，北京：中国建筑工业出版社，2002

3.2.3 Summary

In traditional real estate projects,especially in those high rises, residents seem to be imprisoned in their apartments lacking communication with the neighborhood. On one hand it is due to the modern living models that large numbers of people are gathered in one building and the individual is lost in the crowds. On the other hand as Jan Gehl said in “Life between Buildings” that it was found the physical environment can influence the social status of the residents in different levels and it can be designed to block or invite different forms of communication [1]. Although we can find lots of public space in the modern living communities, but in most cases unlike the Co-housing projects they are not used very well and little communication is established among the neighbors. Besides due to the consideration of cost, traditional housing projects usually only adopt several standardized units while in Co-housing projects by deliberately consideration sometimes even each unit, with skillful design, has its own customized plan which is not at the cost of much money. (Fig.3.37)

In this part the author will summarize the strategies Co-housing adopting to realize individualized life in two aspects: Living Style and Living Space.

On the aspect of individualized living style, what Co-housing projects usually seek for can be summarized as two categories (Fig.3.38):

(1) **Defined Requiremnets on the Residence Composition**: this type of Co-housing will define the composition of the future residents. This process can happen in the group forming period or the new member recruiting period. The forms of this type contain retirement house type, single woman apartment type, multigeneration housing type and other

1 Jan, G. (1987) Life between Buildings, New York: Van Nostrand Reinhold Company Inc.

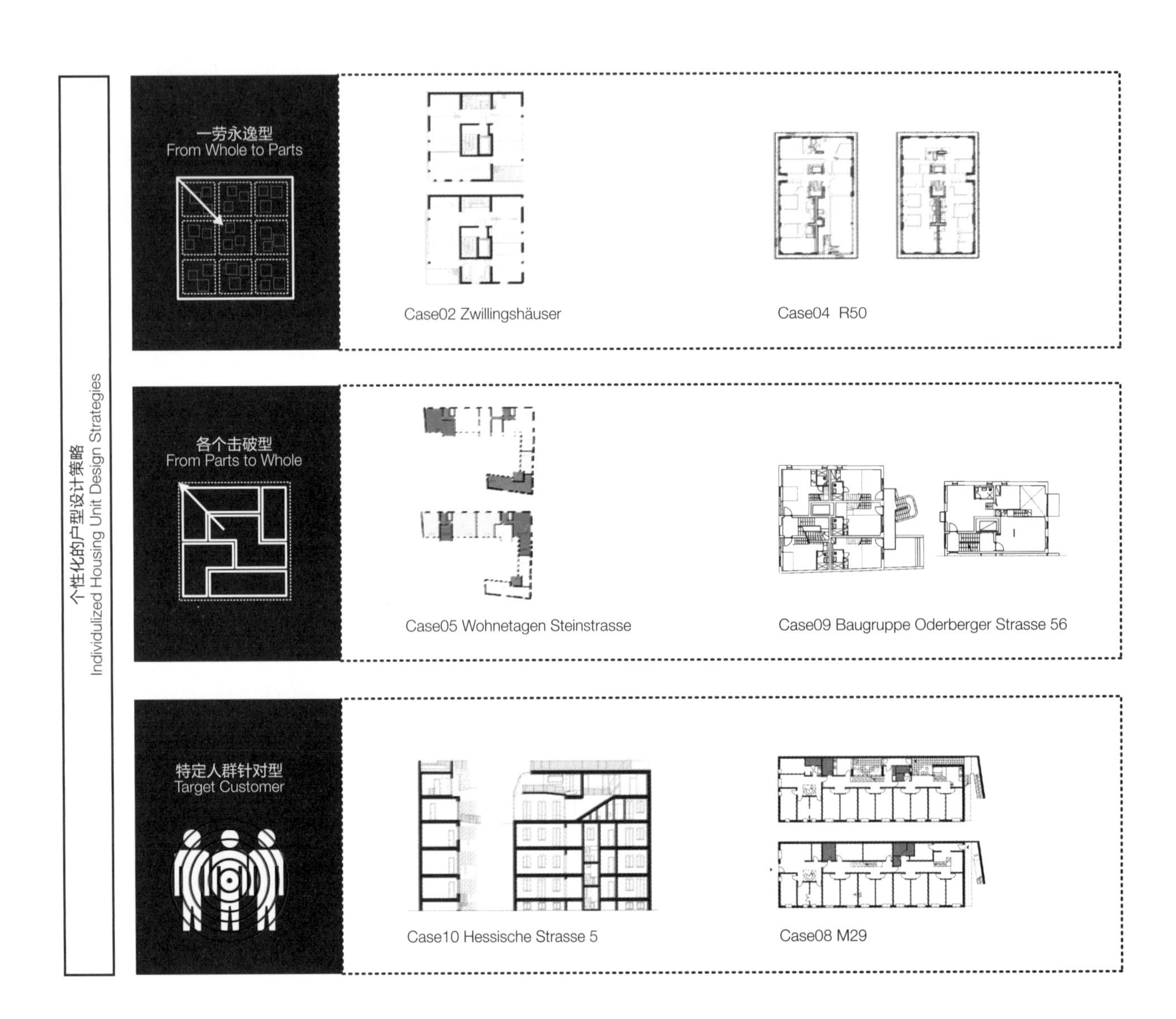

Fig.3.39个性化户型设计策略 Individualized housing unit design strategy

每一户设计单独的户型并不难，难在这些户型必须要相互组织在一起，形成一个完整的建筑。联建住宅居住空间的设计难点有二，一是如何以有限的资源与成本最大化满足各成员的需求，二是如何组织各个性空间。

对于第一个难点，通过对具体案例的研究总结，主要有三类解决策略（Fig.3.39）：

（1）**一劳永逸型**：设计先从整体再到局部，综合各方面的意愿，先行设计整体构架与居住单元的大致分隔，然后针对单元设计一系列模块化的改造构件，住户根据自己喜好，通过相互搭配产生不同组合。一般这种情况，户型的灵活度较高（自由的平面），能够根据特定需要进行再次分隔改变，并且户型中都预留了扩展的构件（门洞、窗、移墙）。这种策略使得建筑能够被反复利用，在建筑使用方进行更替时适应性强。

（2）**各个击破型**：设计先从局部再到整体，建筑设计师针对每户的需求进行设计，然后综合考虑所有户型进行整合，设计成本较大，但高度贴合住户需求，不过在涉及建筑使用方更替时，其适应性不及第一种策略。

（3）**特定人群针对型**：联建住宅项目中还存在着这样一种设计方式，即建筑师针对特定假想人群进行设计，然后再去寻找合适的人群结成联建团体。通过这样设计出来的住宅通常是非常规的，可能是根据某些特殊地形设计的非常规户型，也可能是倡导某种集体宿舍式居住模式的户型。这种类型更具实验性与先锋性。

对于第二个难点如何组织各居住单元，主要有以下几种方式（Fig.3.40）：

（1）**均匀排布型**：这种类型适用于户型变化不大的项目，按照传统方式线性均匀排布，有时亦会插入次一级的小型共享空间。

（2）**平面交错型**：这种类型主要是指户型在同一平面上产生交错，但不涉及垂直向的耦合，通过叠合的方式共同

types that show the concern to the vulnerable groups.

(2) **Defined Requirements on living models**: this type of Co-housing has no special requirements for its membership composition but explores alternative living models. The forms of this type contain collective dormitory type, function-mixed complex type (mixture of residence, office, culture and commerce).

Since there are many parties in one Co-housing group, each party has its own likes and dislikes. Based on the rules of Equity and Equality, the final project plan is the result of compromise which all the parties are agreed to and satisfied with. To design a housing unit is not difficult but how to integrate different housing units together to form a whole building is the most difficult part. When the architects are dealing with this difficulty, two aspects are involved: one is how to realize the clients' demands to the most with limited resources and budgets; the other is how to organise the individualized units.

For the first aspect, based on the selected cases, there are usually three solutions(Fig. 3.39):

(1) **Design from whole to parts**: firstly the designers integrate wishes of all parties and design the frame of the building and the general division of living units. Then they developed a series of modular components including partition walls, windows & doors and so on and each type has several options. Residents can choose the components according to their willing to divide the units thus creating different individualized housing units. When the users change in the future, the layout of the room can be flexibly changed without many efforts.

(2) **Design from parts to whole**: firstly the designers design the individual unit according to each user, and

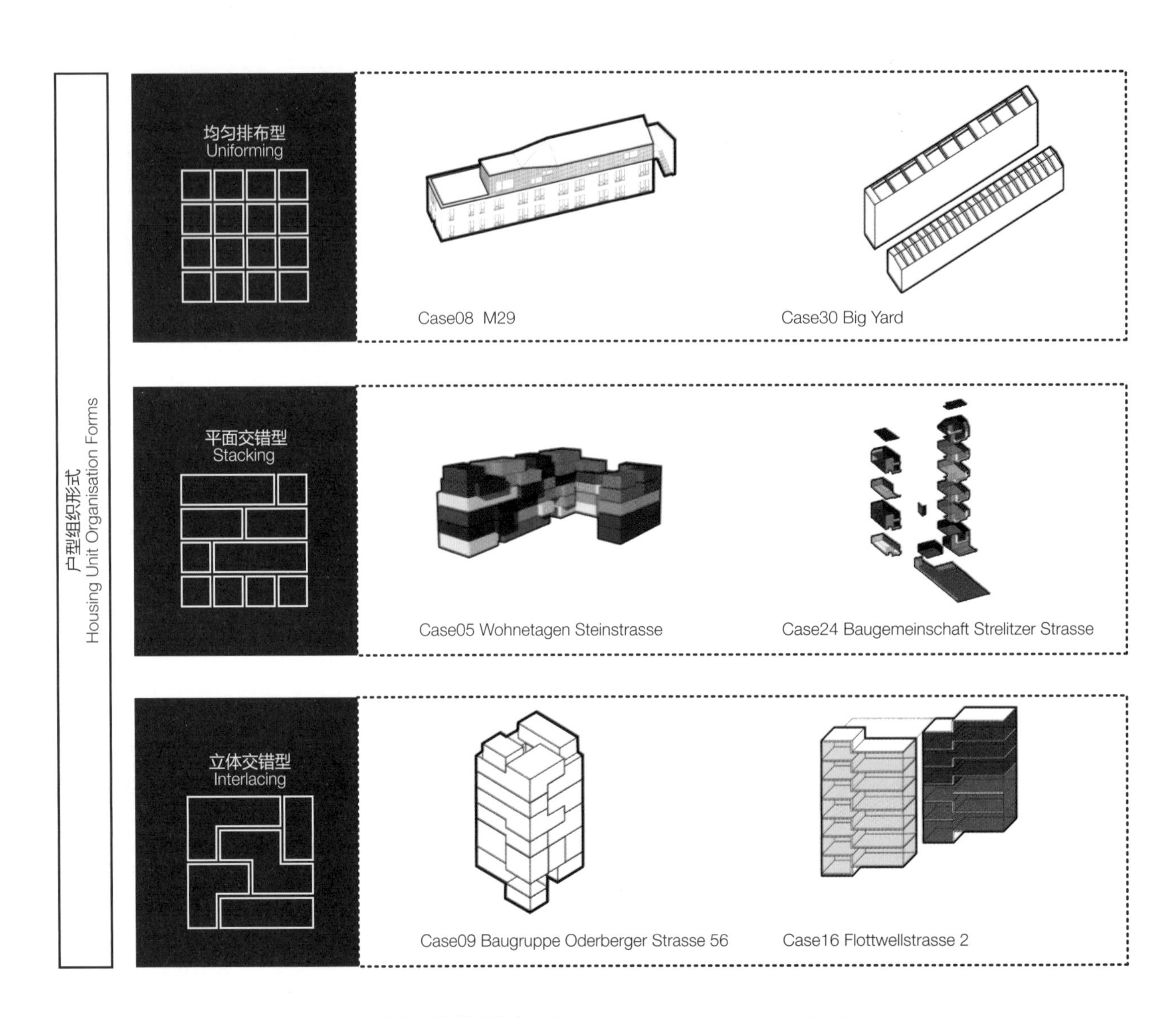

Fig.3.40个性化户型组织形式 Individualized housing unit Organization forms

构成建筑整体，户型个性化程度大于前种类型。

（3）**立体交错型**: 在这种类型中，户型个性化程度较大，相互之间通过耦合共同构成建筑整体。

integrates the non-standardized units together to form the whole building. Usually the design cost and construction cost of this strategy is higher than the former one, but it is highly fit the users. When the users change in the future, it has much less adaptability than the former one.

(3) **Design for target group**: firstly the designers design the building for presumed group and then find these target group to initiate the projects. Usually in this situation it will develop lots of unusual housing units even some “extreme units” . It is more experimental than the former two strategies.

As to the second aspect: how to organize the living units, it majorly contains three ways (Fig.3.40):

(1) **Uniforming Type**: this type is used in the projects that have standardized units. The living units are linearly arranged and sometimes shared spaces are inserted into them.

(2) **Stacking Type**: in this type usually the units interlace with each other on the same floor. Through stacking different layers form the whole building. It is more individualized than the former one.

(3) **Interlacing Type**: in this type, different units interlace with each other both vertically and horizontally. Like “Toy Bricks” they form the whole building. It is more individualized than the former two. It is often used in space-intensive buildings.

传统住宅
用材与结构较保守
Traditional Housing
conservative materials and structure

联建住宅
对于材料与构造有实验性尝试
Co-Housing
innovative experiment on new materials and structure

传统住宅
孤岛，有些内部可持续性好
Traditional Housing
Isolated Isle， some have good sustainability

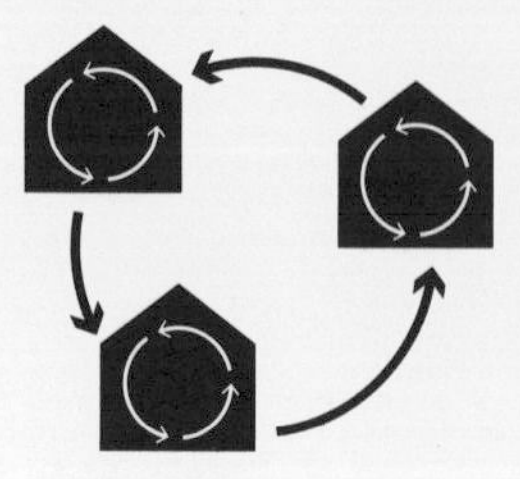

联建住宅
存在联系，内部可持续性好
Co-Housing
Network，have good sustainability

Ecology
生态性

need to make their living area environment-friendly and keep resource sustainable for both economic and ecologic considerations

出于生态与经济方面考量，构建环境友好型生活空间，确保资源、能源的可持续利用。

Sustainability
可持续性

Traditional Housing

Conservative materials and structure / Isolated Isle，some have good sustainability

Co-Housing

Innovative experiment on new materials and structure /Network，have good sustainability

传统住宅：用材与结构较保守。孤岛，有些内部可持续性好
联建住宅：对于材料与构造有实验性尝试。彼此存在联系，内部可持续性好

Fig.3.41生态性与可持续性 Ecology & Sustainability

3.3 生态·持续

Stadtbau GmbH负责人Constance Cremer曾提到"想要建设一栋联建住宅会引发很多问题。从寻找合作伙伴，选择组织形式到规划和建设事宜，方方面面的问题都迫切需要得到解答。……长期的可负担性、居住环境、未来可持续性将是问题所在"[1]。不可否认的是，柏林联建住宅项目正引领着建筑界的生态风潮，虽然前期和中期投入的成本相对较高，但是从长远的角度来考虑还是十分经济的，所以绝大多数柏林联建住宅项目都在实现与深化着生态建筑标准和生态可持续的生活方式。联建住宅项目有很多都是先锋实验项目，其中，新技术实行的是严格于现行节能建筑标准的新标准。在未来，这些项目将在资源利用上扮演重要的示范性角色。据相关研究表示，建筑所需能源的15%到50%取决于使用者的行为习惯，不同于传统的居住模式，联建住宅项目可以最大限度地挖掘节能的可能性（Fig.3.41）。

3.3.1 生态地建造：可持续的建材与建构

柏林联建住宅项目从设计之初就想竭力实现生态、经济、社会等多方面的可持续发展。联建住宅项目源源不断的创意反映在其建筑品质之上，反映在联建团体以及他们所处社区中承担的重要角色之上，同时也反映在精心设计的生态建筑之上。很多联建住宅项目都在积极探索建筑建造过程中材料与建构方式的更多可能性。

1 作者译自 Ring, K. (2013) Selfmade City Berlin: Stadtgestaltung und Wohnprojekte in Eigeninititie, Berlin: JOVIS，148

3.3 Ecology & Sustainability

"Thinking about founding a communal building or Co-housing project raises many questions. Answers are needed for questions ranging from the search for fellow participants and the organizational and legal layout to the entire matter of planning and building....Long-term affordability, residential environment and, where required, models for the future can be an issue.[1]"(Cremer [2],2013) There is no doubt that Co-housing projects are at the forefront of exploration to Eco-buildings. Although the preliminary input is relatively higher than the normal housing, Eco-housing can be quite economical in the long run, so many Co-housing projects in Berlin have practiced and optimized the Eco-sustainable building standards. In the future these projects will play demonstrative roles in the resource utilization. According to statistics, 15%-50% energy consumption of the building is up to the users' behavioral habit. Unlike the traditional living styles in the normal housing projects, living styles in Co-housing projects can help to explore the potential of energy-saving in everyday life (Fig.3.41).

3.3.1 Ecological Construction: Sustainable Material and Structure

The ambition to realize the sustainability on Ecology, Economy, Society and other aspects goes along with the development of Co-housing projects. Their innovation not only reflects on the qualities and forms of the projects, but also on the deliberately designed ECo-housing. Many Co-housing projects are searching for the new possibilities of materials and structures on the process of construction.

1 Source: Ring, K. (2013) Selfmade City Berlin: Stadtgestaltung und Wohnprojekte in Eigeninititie, Berlin: JOVIS: 148

2 Constance Cremer, Stadtbau GmbH, Berlin

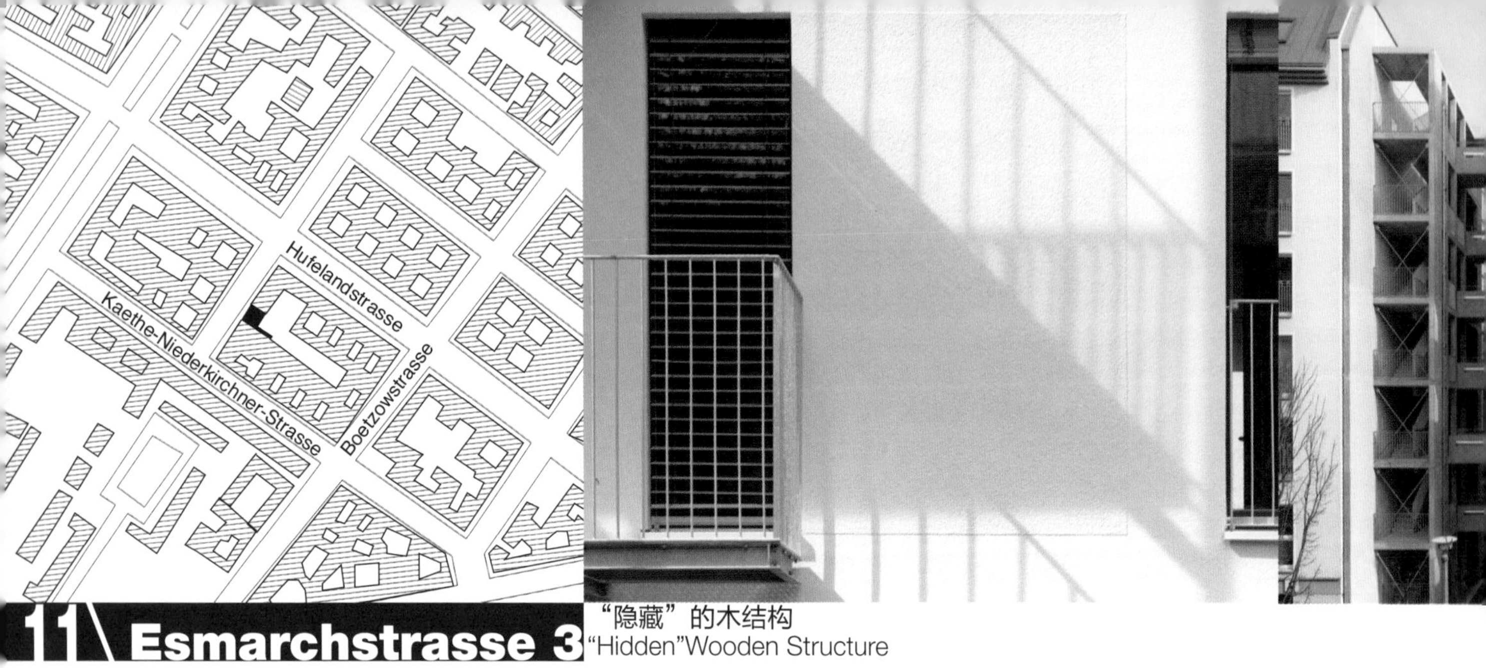

11\ Esmarchstrasse 3

“隐藏”的木结构
“Hidden”Wooden Structure

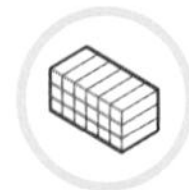 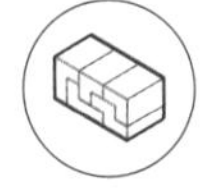 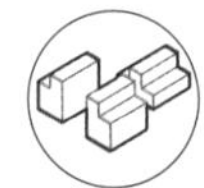

 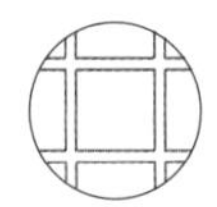 

项目概况 Project Profile

项目名称 Project Name：	Esmarchstrasse 3
项目地址 Project Address：	Esmarchstrasse 3
建筑设计师 Architect：	KADEN-KLINGBEIL ARCHITEKTEN
基地面积 Site Area：	N/A
建筑面积 Total Area：	1270 m^2
项目成立时间 Founding Time：	2006
项目状态 Project Status：	完工 finished

Fig.3.42项目区位及实景 Project location & scene

E3 项目为多层木结构树立了全新的范例，在国际上被广泛宣传。该项目位于 Esmarchstrasse 大街 (Fig.3.42)，通过外立面人们无法识别出它是一栋木构建筑，而这正是设计师想要达到的效果。

“在德国只允许在市中心建造三层高的木结构建筑，在 2002 年修改了建造法案之后（柏林在 2006 年二月份才调整），对城市木构建筑放宽了限制，出地平面高度限制为 13m，更高的建筑仍然必须采用不燃材料。但是就在苛刻的柏林建筑法案中，仍然留有余地，即 § 27 (1) 和 § 31 (1)：关键性构件与天花板都可不必是防火材料，只需要高度阻燃即可，这也是通过长期的协商所争取的，建筑师援引这两条，第一次提出将木材作为七层高城市的建筑材料”[1]。建筑方案得以通过主要在于其防火理念，而这在设计阶段就已经成形。由预应力混凝土制成的开放性楼梯间紧靠北侧防火墙设置，通过桥与每层住户相连，将楼梯间与主体脱离，从而形成了一个开口（Fig.3.43）。正是独立式的楼梯间使得消防部门确信在开放的疏散路径上是不会产生大量烟尘的。此外，每层楼和核心筒上都安装了消防探测器，覆盖了所有的路径，在日常生活中稳定地发挥着重要作用。

1 作者译自 Kleinein, D. (2008) ‘Esmarchstraße 3: Kritische Verkapselung’, Bauwelt, 18/04, 18

Project E3 is a multi-storey wooden structure housing located in Esmarchstrasse 3(Fig.3.42). From its chess-like punched façade there is no hint showing it is made of wood, which is the intention of the designers.

“In Germany only 3-storey wooden structure buildings can be constructed in the city center. After Construction Act revised in 2002 (Berlin until 2006), the restriction is relaxed and the construction height limit is 13m up the ground. If the building is higher than the limit, it should adopt nonflammable materials. In the more strict Berlin Building Act, there are still some regulations can be discussed like § 27 (1) and § 31 (1): key components and ceiling boards have no need to be fireproof materials but flame-resistant materials. With these two regulations the designers negotiated with the relevant department and proposed to use wood as the major material for the 7-storey urban housing[1].” The design was approved by the relevant department because of its fire-protection concept. The open escape staircase made of prestressed concrete is set next to the north firewall and connected to each floor by

1 Translated by the author from source: Kleinein, D. (2008) ‘Esmarchstraße 3: Kritische Verkapselung’, Bauwelt, 18/04, 18.

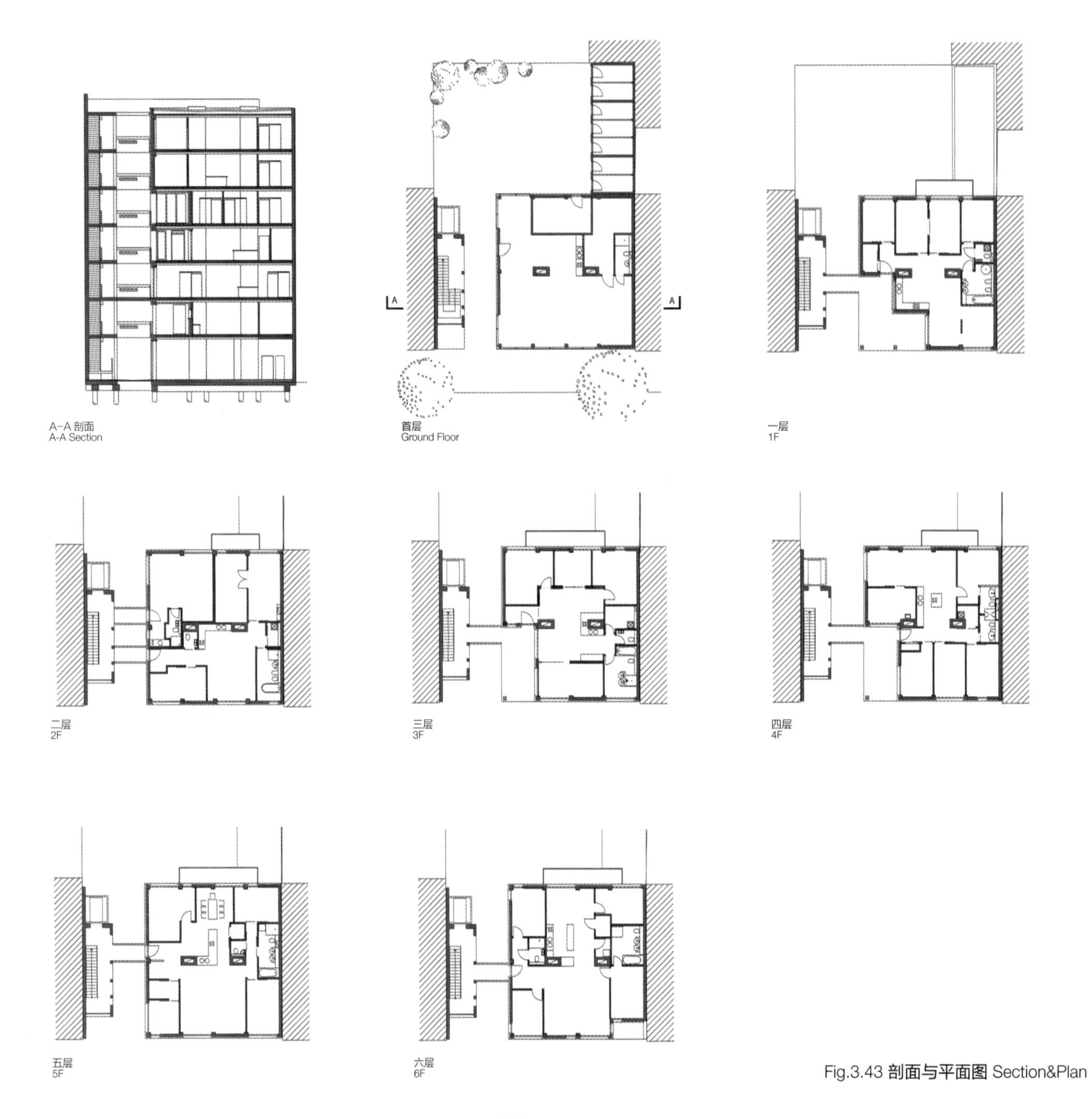

Fig.3.43 剖面与平面图 Section&Plan

整个建筑只有在天花板的底面才能看到大面积的建筑实木，托梁与实木优雅地结合在一起，工程细节做得十分到位。实际上，这栋建筑并非出于审美需求才选用木材，而是在探索一种替代混凝土的、新的建造方式。

"bridges"(Fig.3.43). It is this independent escape staircase that persuaded fire department that it won't generate lots of smokes in the open path when it is on fire. In addition every possible paths are within the working zones of the fire detectors.

It is not out of the aesthetic demand that led to the use of wood, but out of the willing to work out an alternative and sustainable construction materials and methods to build multi-storey housing in the city center.

12\ 3Xgrün

可以复制的木构体系
Duplicable Wooden Structure System

 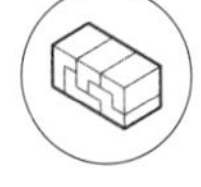 

 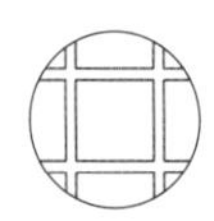 

项目概况 Project Profile

项目名称 Project Name：3Xgrün
项目地址 Project Address：Görschstrasse 48/49
建筑设计师 Architect：ARGE ATELIER PK, ROEDIG.SCHOP, INSTITUT FUER URBANEN HOLZBAU (IFUH)
基地面积 Site Area：1340 m^2
建筑面积 Total Area：2877 m^2
项目成立时间 Founding Time：2009
项目状态 Project Status：完工 finished

Fig.3.44项目区位及实景 Project location & scene

本案例是木构建筑的另一尝试，设计理念与上一案例有很大区别。前文提到在德国建造木构住宅有很多限制，比如建筑限高，修订过的建筑法规所允许建筑最高出地平面13m，这相当于五层建筑的高度。木材制品灵活度相当高，可以用于多种建筑类型，但是当时仍然缺乏适宜的多层木构住宅概念，用以充分利用预制木构件及其建构体系的优势。基于此背景，由 fertighauscity5+ [1] 领队的研究小组发展了一整套城市木构建筑体系（Fig.3.45），适应各种大小的地块、各种建筑体块。

从研究一开始，木构的策略作为全局概念就已确定下来。在规划阶段，设计师充分考虑了各种使用人群对于未来居住空间的使用需求 (Fig.3.46)。在此基础上，研究小组制定了客户订制化的实施模型，其系统化的规划过程简化了用户的决策过程。此外，高度先进的技术生产过程使得个性化的平面与立面得以实现，预制化的构件能够十分迅速与准确地在现场完成安装。

3XGRÜN 项目是第一个依据 fertighauscity5+ 研究成果实施的预制木构多层住宅实例。项目名称中的“3”意指该居住社区拥有花园、阳台与共享的屋顶露台这三类充

1 该机构主要从事对城市多层木构建筑的类型学和技术方面的分析，并强调预制和使用者参与。

3Xgrün （Fig.3.44）is another attempt of wooden structure housing. With the different design concept from project E3, it is still worth of discussing. Wooden structure buildings have high flexibility and can be applied in many architecture types. During that period there was still lacking appropriate multi-strorey wooden housing concept that fully makes use of advantages of premade wooden components and structure systems. Under this background, the research team led by fertighauscity5+ [1] developed a set of urban wooden structure building system (Fig.3.45) to adapt to different sizes of the sites and different types of architectures.

3XGRÜN is the first finished multi-storey wooden structure housing based on the research of fertighauscity5+. At the beginning of the project, wooden structure strategy was decided and different user groups' demands on future living spaces were also considered. Based on it the research team defined the customization implementation model that helps simplify the process of users' decision making(Fig.3.46).

Pre-stressed concrete is used in the positions which are

1 This research team works on the typological and technical analysis to multi-storey wood construction in urban areas in the light of prefabrication and user participation.

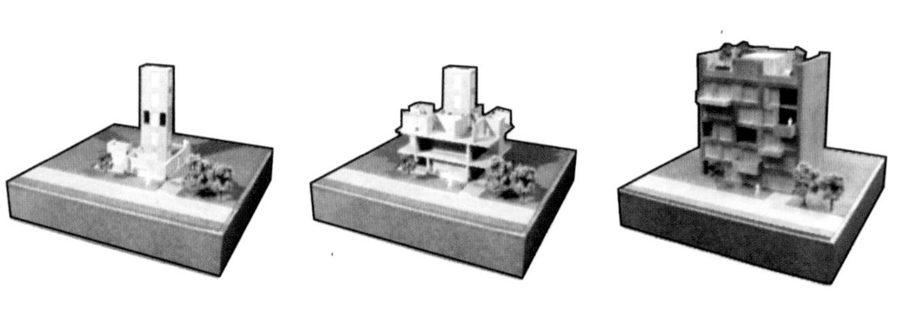

Fig.3.45木构研究模型 Wood structure research model

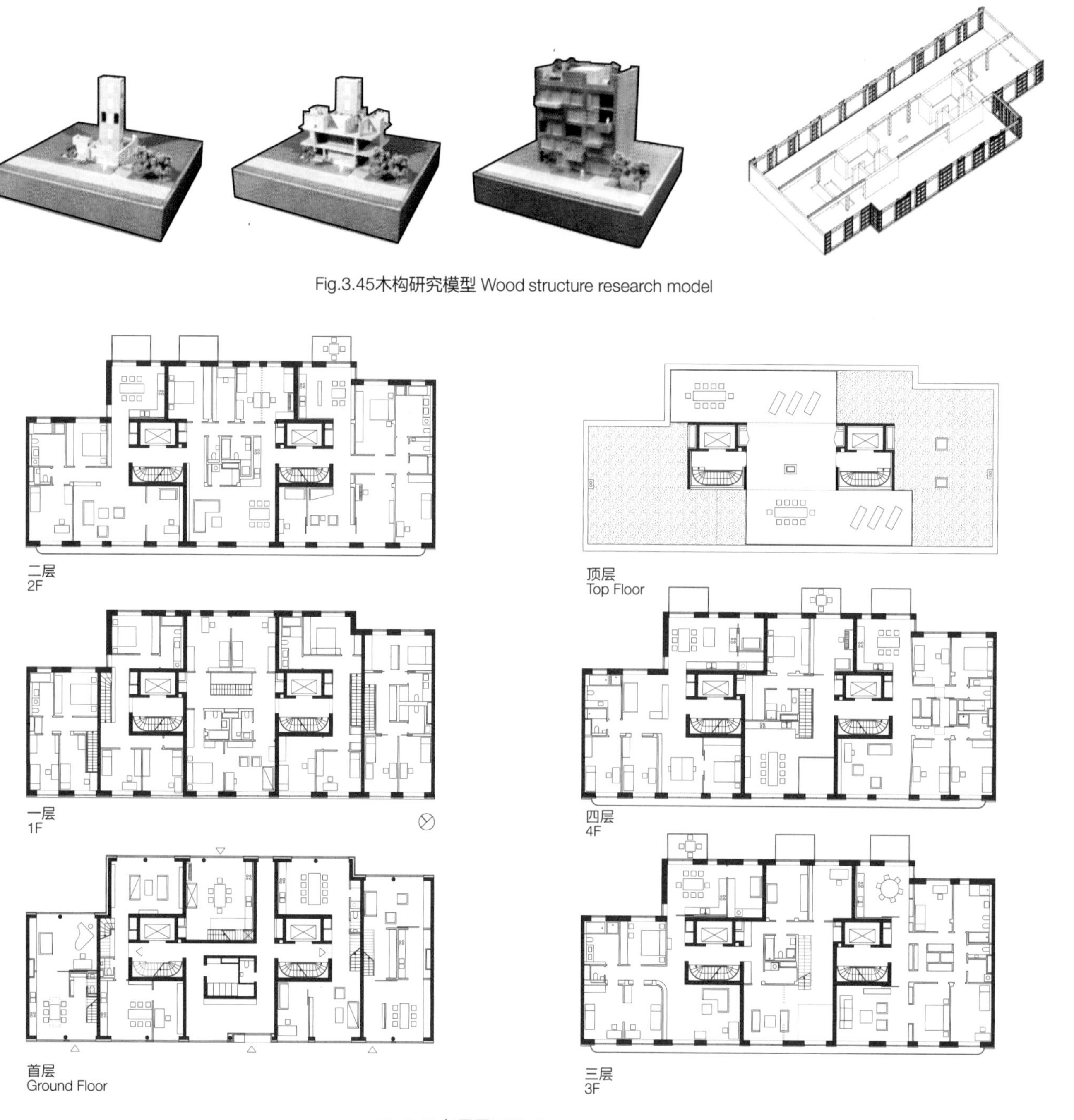

Fig.3.46 各层平面图 Each floor plan

满“绿意”的空间。本项目坐落于 Görschstraße 48 号和 49 号 (Fig.3.44)，在边界范围内，综合考虑了周边既有建筑的檐口高度、建筑轮廓以及前庭花园的结构。

除了荷载与湿气较大部位如基础、地下室、防火墙和楼梯间由预应力混凝土制成，其他部位均采用木构件。所有木构件都预制成了大尺寸的元件。天花板材系统、墙构件在当地的木材加工厂完成的同时，预应力混凝土墙也是作为半成品预制和运输的。在此过程中 450m^2 的地下室仅仅用了两周的时间完成。本案例充分展示了预制木构建筑的优势与可行性。

under great pressure and humidity like the foundation, basement, firewall and the staircase. Ceiling and wall components were manufactured in the local timer-processing mill; prestressed concrete walls were transported as half made components. This case demonstrates fully the feasibility and advantages of premade wooden structure architectures.

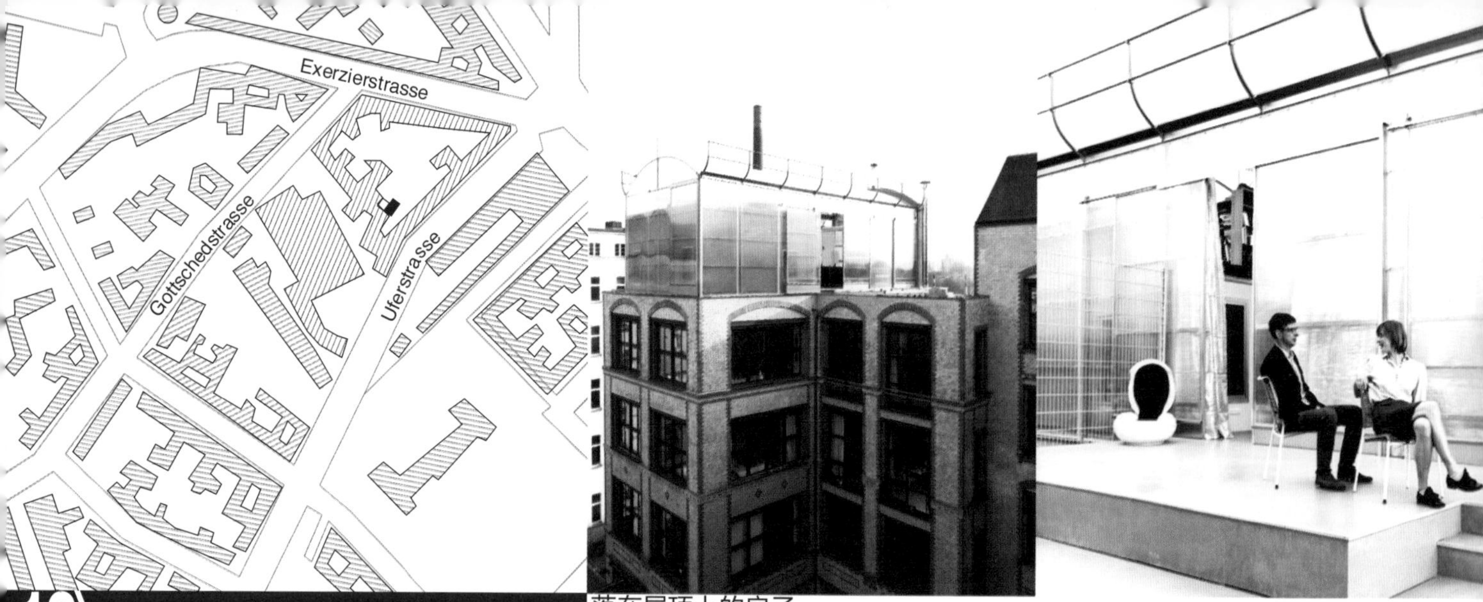

13\Hegemonietempel 落在屋顶上的房子 A House landing on the Roof

 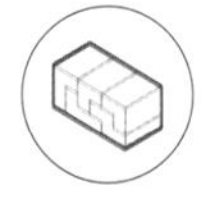 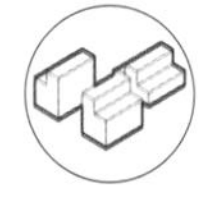

 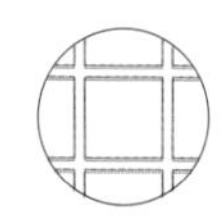 

项目概况 Project Profile

项目名称 Project Name： Hegemonietempel
项目地址 Project Address： Uferstrasse 6
建筑设计师 Architect： CHRISTOF MAYER, BUERO FUER ARCHITEKTUR UND STAEDTEBAU
基地面积 Site Area： 283 m^2
建筑面积 Total Area： 134 m^2
项目成立时间 Founding Time： 2009
项目状态 Project Status： 完工 finished

Fig.3.47项目区位及实景 Project location & scene

本项目选址在一栋顶层被炸毁的老旧工厂之上，整个建筑看起来就像是一个温室。面对目前全球关于气候保护与生态建造的问题，本项目给出了自己的解答 (Fig.3.47)。

该项目作为联建住宅比较特殊的地方是它的团体成员只有两名，他们都是艺术家，在 2007 年卡塞尔文献展上他们看到了一名法国建筑师关于温室的展览，由此萌生出利用这种结构材料建造房屋的想法。由于结构比较轻，故而基地的选择限定在了屋顶这一未被“开发”的空间类型上，之后在经过与建设部门长达 14 个月的协商（关于防火、安全疏散等方面的要求），项目终于开始实施，并于 2010 年正式完工。整个建筑由三部分组成（Fig.3.48），一是作为围护的温室结构，另外就是两组砖砌的一层半高固定模块（有供暖设备），主要用于卧室、书房以及厨房卫生间等功能，这两个模块之间的区域则为自由布置区，该区域进行自然采光。整个建筑一年的用地租金为 2700 欧，建筑成本为 70000 欧，可以说是一次低造价住宅的尝试。

Hegemonietempel is located on the roof of an old factory whose top floor was ruined in the WWII(Fig.3.47).

The Whole Story began like this: the two owners visited Kassel Documenta in 2007, and saw a temporary pavilion built by a French architect with conservatory material, and they got inspired and decided to build a house like the green house. Because of the light structure, the site is defined within the roof areas. Then they found this old industrial plants in good condition: the top was destroyed by the bomb in World War II, but the original staircase and freight elevators can still be accessible directly to the roof. After 14 months of negotiations with the construction department, finally the project passed the harsh famous German standard indicators of the inspection and was finished in 2010.

The house consists of 3 parts (Fig.3.48): green house covering and two one and a half-storey brick rooms (heated). The two rooms are used as bedroom, study, kitchen and toilet. Between them is the free zone with natural lighting and its function is multiple. The cost of the house is 70000 euro and it is a low-cost and low-tech housing attempt.

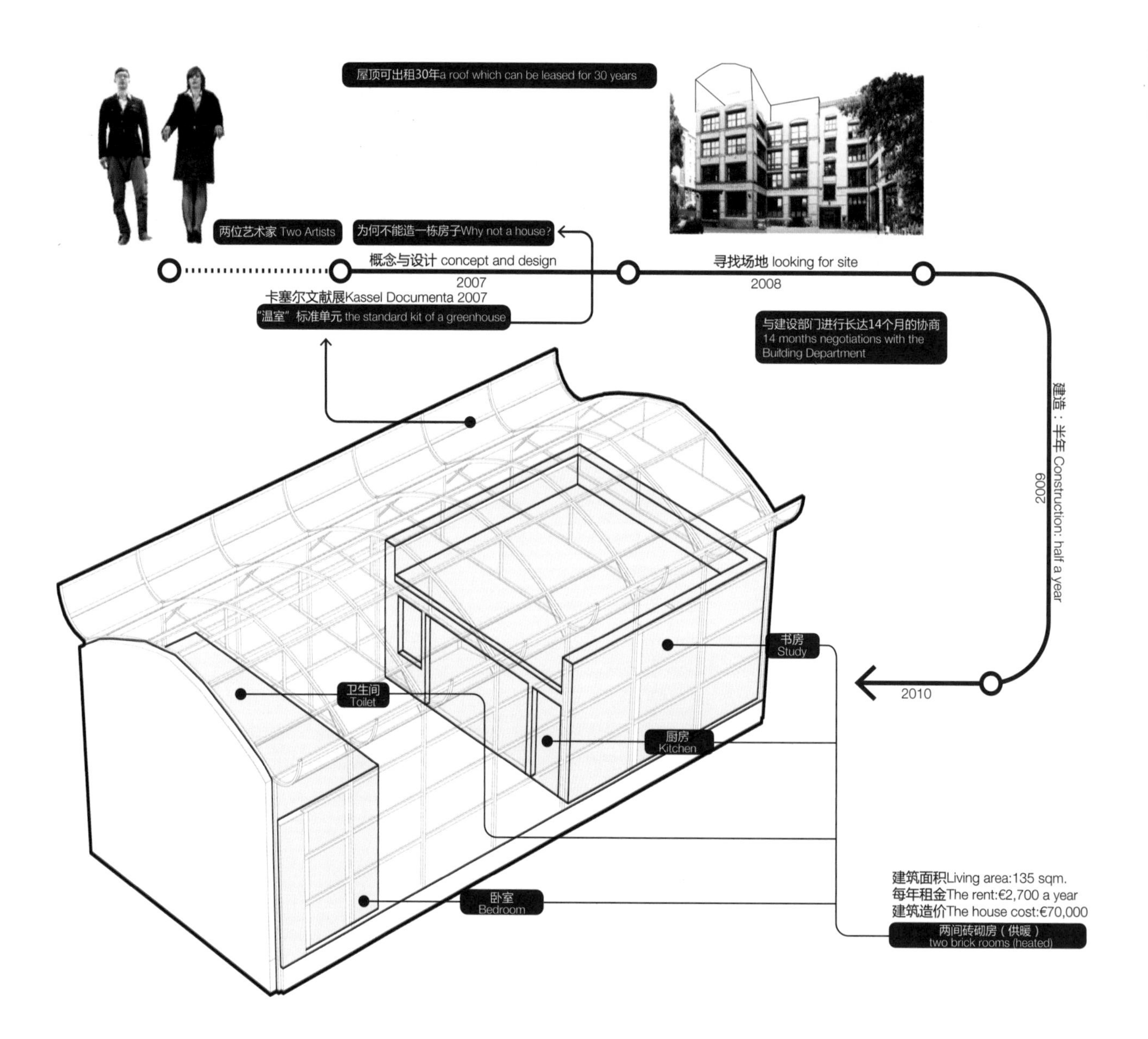

Fig.3.48 **项目发展及其建构** Project's development and its tectonic

3.3.2 生态地居住：可持续的生活方式

不管是出于长期经济可负担性的考虑，还是为了倡导健康生态的行为习惯，大部分联建住宅项目都是高品质的节能生态住宅，其中有主动式也有被动式。居民的生活方式也是可持续的，这种可持续性体现在居民生活所影响的自然环境，以及社会与经济方面。

3.3.2 Ecological Life: Sustainable Living Style

Not only out of the long-term affordability consideration but also out of advocation to the sustainable healthy living habit, most Co-housing projects are Eco sustainable-housing with high quality. Correspondingly the Co-housing residents practice a sustainable living style.

被动式住宅项目
Passive Housing

 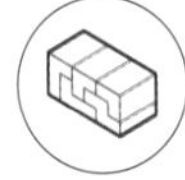 

 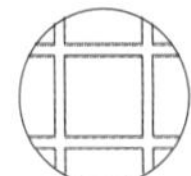 

项目概况 Project Profile

项目名称 Project Name：	Living in Urban Units
项目地址 Project Address：	Schönholzer Str.13-14
建筑设计师 Architect：	DEIMEL OELSCHLAEGER ARCHITEKTEN PLAANPOOL ARCH. MIT WERKGRUPPE KLEINMACHNOW
基地面积 Site Area：	857 m^2
建筑面积 Total Area：	2610 m^2
项目成立时间 Founding Time：	2006
项目状态 Project Status：	完工 finished

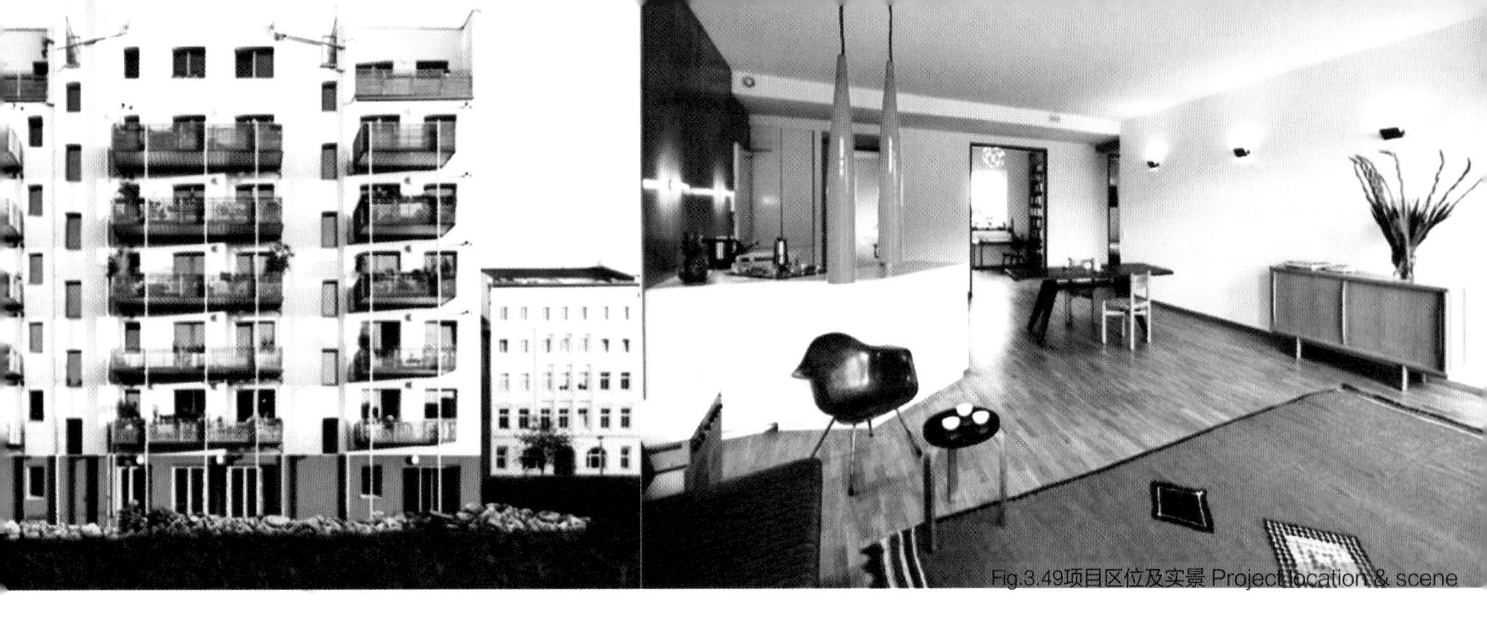

Fig.3.49项目区位及实景 Project location & scene

该联建团体由青年与老年、单亲家庭与普通家庭所组成，他们共同建造了这栋充满创意的被动式节能住宅[1]。项目由 20 户大小各异的居住单元构成，希望实现居民的混合，在社区生活的氛围中互相扶持。每户居民生活在独立的套间之内，通过一系列的共享空间（底层的共享庭院、共享活动室以及共享的屋顶露台）进行交流 (Fig.3.49)。

被动式住宅利用室内热源以及太阳能设备进行热能循环并且在封闭的室内系统里避免热能散失。可控通风系统减少了通过自然通风所散失的热量，配合热回收系统在冬天持续不断地往室内输送预热的新鲜空气。如果有必要，还可以补充安置加热设备。按照被动式住宅的建造标准，平均每栋住宅的热能消耗量可以降至十分之一。

根据达姆施塔特被动式住宅研究所标准，该建筑供暖所需能源为 15kWh/m^{2}a，而德国节能法案标准基础能耗是 34kWh/m^{2}a。所以该建筑比普通建筑大概少用40% 的能源。

1 被动式住宅起源于上世纪 90 年代的德国法兰克福。这类住宅主要通过构造做法达到高保温隔热性，并利用太阳能和家电设备散热为居室提供热源，减少或不使用主动供应的能源，即便需要其他能源，也尽量采用清洁可再生能源，房子密封几乎没有任何热量散失在德国。该类住宅成本仅比传统房屋高出 5%–7%。（http://www.baike.com/wiki/%E8%A2%AB%E5%8A%A8%E5%BC%8F%E4%BD%8F%E5%AE%85）

The Co-housing group consists of young and old, single family and nuclear family, it aims to create an innovative passive house[1]. The building is made up of 20 different living units. Each party lives in the separate apartment and shares a series of communal space (shared courtyard, communal activity room and shared roof terrace)(Fig.3.49).

The principle of pasive house is to utilize the indoor heat sources and the solar collectors to go on thermal cycle and avoid heat loses in the closed indoor system. Controlled ventilation system can reduce the heat loses in the natural ventilation and co-work with heat recycle system to send the pre-heated fresh air in to the interior.

The heating requirement for the building is only 15 kWh/(m^{2}a)according to the standards of the Passivhausinstitut in Darmstadt (PHPP), and the primary energy need is 34 kWh/(m^{2}a) according to EnEV(German energy saving act standards). The building needs approximately 40 percent less energy than a standard building.

1 Passive house (Passivhaus in German) refers to a rigorous, voluntary standard for energy efficiency in a building, reducing its ecological footprint. It results in ultra-low energy buildings that require little energy for space heating or cooling. Although it is mostly applied to new buildings, it has also been used for refurbishments. The first Passivhaus residences were built in Darmstadt, Germany in 1990, and occupied by the clients the following year. (Source: http://en.wikipedia.org/wiki/Passive_house#cite_note-NYT-2010.09.25-1)

06\Malmoeer strasse29

经济、资源的可持续方式探索
Exploration on Economy and Resource Sustainability

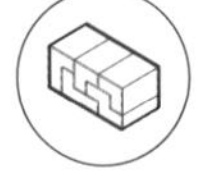
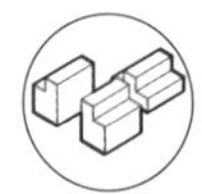

项目概况 Project Profile

项目名称 Project Name： M29
项目地址 Project Address： Malmoeer strasse 29
建筑设计师 Architect： CLEMENS KRUG ARCHITEKTEN
基地面积 Site Area： 800 m^2
建筑面积 Total Area： 1012 m^2
项目成立时间 Founding Time： 2008
项目状态 Project Status： 完工 finished

Fig.3.50项目区位及实景 Project location & scene

可持续化的生活方式不仅仅局限于项目内部的能源利用，还可以通过各种组织联系在一起，它们共享资源，在更大的范围内实现资源、经济的可持续性。这里涉及到市民经济（Civic Economy[1]）的概念。市民经济架起了公众、个体与第三方之间的桥梁，通过一些创造性的方式使多方进行合作，更高效地整合利用各方资源，使得市民成为共同生产者和投资者，而非仅仅是资源的消费者。

比较有代表性的例子是 M29 项目 (Fig.3.50)，它是 Wilde Gaertnerei 项目的成员，Wilde Gaertnerei 主要从事基于社区的农业项目，位于柏林以北 30km，为许多联建项目、当地市场、餐厅提供新鲜的蔬菜。Wilde Gaertnerei 项目可以借用 M29 项目的共享空间在当地进行宣传，而 M29 项目可以以优惠的价格购买新鲜的蔬菜水果。此外，M29 项目又跟当地的发电厂有合作，也能以优惠的价格购买电力资源，同时它又是出租房合作社 (Mietshaeuser Syndikat) 的成员，能够以较低的价格租赁项目场地和建筑。总之，很多联建项目往往处在多个组织之中，通过组织，他们共享资源和经验，从而更好地促进项目本身可持续地运营（见 Fig.3.23）。

1　关于市民经济更多案例参见 Ahrensbach, T., Beunderman, J., Fung, A., Johar, I. and Steiner, J. (2011) Compendium for the Civic Economy, London: Calverts Co-operative.

Co-housing projects are not isolated but connected with each other via various social organizations. They share the resources and realize the economy and resource sustainability in a broader scope. Here Civic Economy[1] is involved. Civic Economy bridges the gaps between public, private and third sector. Through innovation civic economy makes possible to unlock and share available resources and to use them more effectively.

Wilde Gaertnerei[2] is an independently founded, since 2012 community supported organic pioneer farm thirty kilometers from Berlin. Project M29(Fig.3.50) is member in it. M29 provides platform for Wilde Gaertnerei to promotion and in return they can get fresh vegetables and fruits at a low price from Wilde Gaertnerei. Besides M29 is in another networks with the local power plant, where they can get power at a reasonable price, and Mietshaeuser Syndikat, where they get the building and site with a low rent (See Fig.3.23). In a word, by extensive networks, Co-housing projects share experience and resources to well sustain their practice.

1 More information refers to Ahrensbach, T., Beunderman, J., Fung, A., Johar, I. and Steiner, J. (2011) Compendium for the Civic Economy, London: Calverts Co-operative.

2 Website: Wildegartnerei.blogspot.de

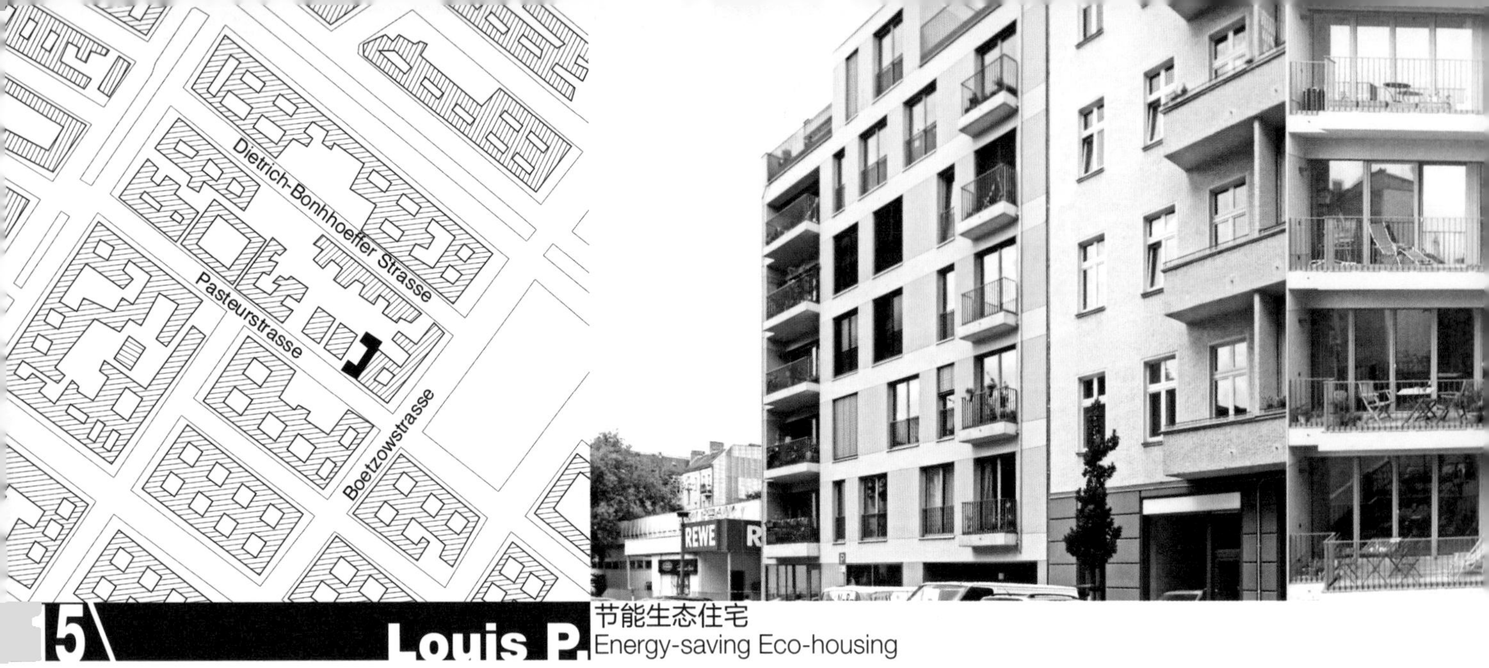

15 Louis P.

节能生态住宅
Energy-saving Eco-housing

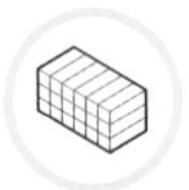 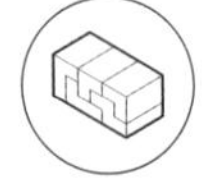 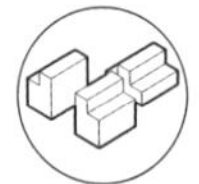

 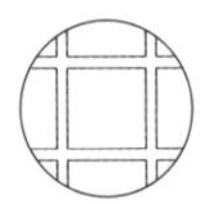 

项目概况 Project Profile

项目名称 Project Name：	Louis P.
项目地址 Project Address：	Pasteurstrasse 27
建筑设计师 Architect：	PLAANPOOL ARCH. MIT WERKGRUPPE KLEINMACHNOW
基地面积 Site Area：	1069 m^2
建筑面积 Total Area：	2543 m^2
项目成立时间 Founding Time：	2008
项目状态 Project Status：	完工 finished

Fig.3.51项目区位及实景 Project location & scene

Louis P. 项目由前后两栋 7 层住宅楼以及连接二者的侧翼建筑组成。项目拥有 20m 进深的内庭院和边院，每个庭院连同屋顶的共享平台都为儿童设置了专门的游乐区域。每户都拥有开敞的阳台、敞廊或是露台，户内净高为 2.8m。对 Louis P. 项目来说，其目标是建立一个拥有大量共享空间的多代居社区。底层容纳了很多灵活性的公共空间：靠近花园的是厨房，浴室，还有一些半室外空间，在这可以举办各种类型的活动 (Fig.3.51)。

整个项目是环保节能节水建筑（Fig.3.52），采用了木屑燃烧供暖系统、太阳能收集系统、热回收通风系统以及中水系统等长期低成本运行的技术。供暖系统的核心是木屑燃烧技术（B）并配有独立的木屑存储空间（C）。公寓通过低温地板加热系统（D）供暖，该系统由余热回收系统补充供暖。所有房间都设置有送风口（E），热交换器（I）通过它们输送预热空气。出风口将厨房、浴室产生的热空气输送给热交换器(I)加热冷空气。在夏天房间由太阳能收集器(A)供应热水。中水循环也是该项目比较重要的的节水理念，通过独立的管网系统（H），生活污水被收集，然后通过中水系统进行生物净化处理后再通过管网（H）用于卫生间马桶冲水或是洗衣机用水。

Project Louis P. is a modern energy-saving multigeneration housing, which consists of two 7-storey buildings and their connection part. The connection part provides each floor with bright and wide corridor spaces. Besides the whole building provides the residents with lots of shared spaces(Fig.3.51).

The key point of this project is its energy and water saving systems (Fig.3.52). It contains the wood pellet heating system, solar collector system, heat recycled ventilation system, gray water system and other long-term low-cost operation systems. The core technology of the heating system is wood pellet heating (B) with an independent pellet storage (C). The apartments are heated by the low-temperature floor heating (D), which is supplemented by the heat-recycled system. Each room is sent in pre-heated fresh air by the heat exchanger (I) via air outlet (E). The hot air generated in kitchen and bathroom will send to the heat exchanger (I) via air inlet (F) to heat the cold fresh air. Water is heated by the solar collectors (A). Gray water system is the important water saving concept of the project. Via separate pipe network (H), domestic sewage is collected. Through the biological purification system it is back to the network (H).

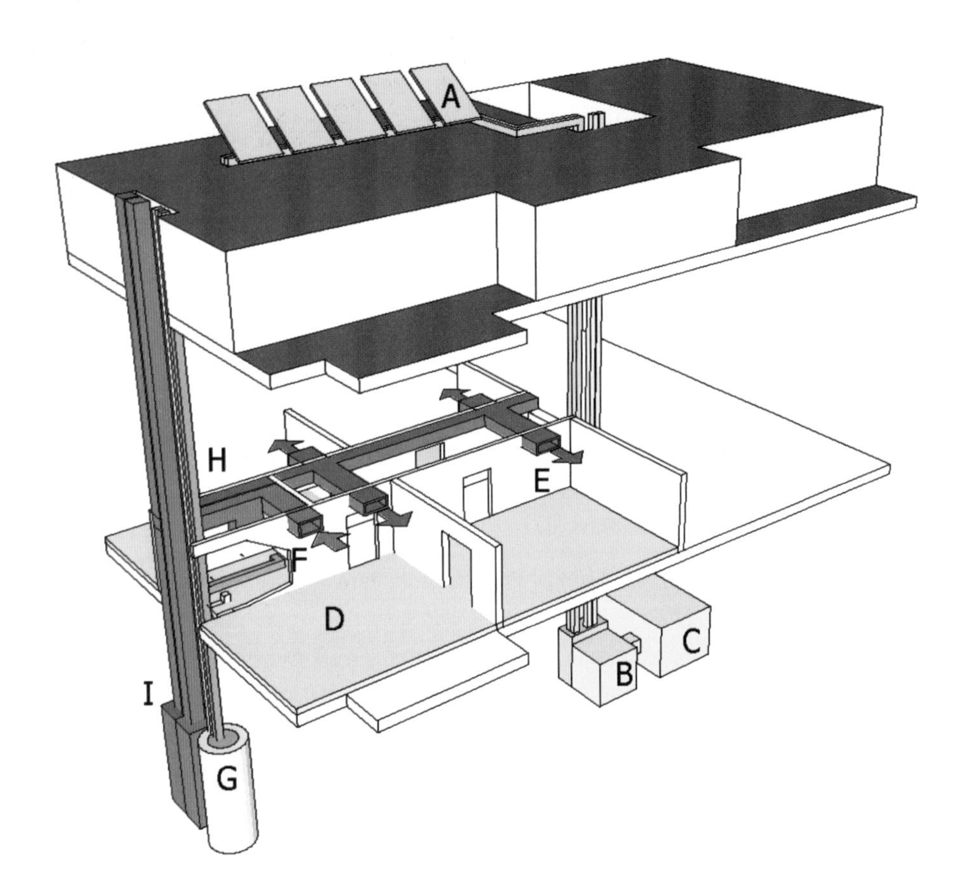

A. 太阳能收集 Solar Collectors

B. 木屑燃烧供暖 Wood Pellet Heating

C. 木屑存储 Pellet Storage

D. 低温地板采暖 Low-temperature Floor Heating

E. 送风口 Air Outlet

F. 进风口 Air Inlet

G. 中水系统 Grey Water System

H. 独立管网 Separate Pipe-network

I. 换热器 Heat Exchanger

Fig.3.52能源利用概念 Energy concept

传统住宅
用材与结构较保守
Traditional Housing
conservative materials and structure

联建住宅
对于材料与构造有实验性尝试
Co-Housing
innovative experiment on new materials and structure

传统住宅
孤岛，有些内部可持续性好
Traditional Housing
Isolated Isle， some have good sustainability

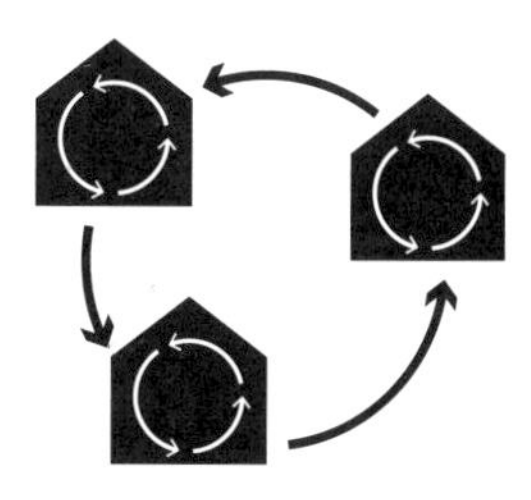

联建住宅
存在联系，内部可持续性好
Co-Housing
Network，have good sustainability

Fig.3.53联建住宅与传统住宅在生态性与可持续上的区别
Difference between Co-housing and traditional housing on Ecology & Sustainability

3.3.3 小结

德国向来拥有最高的生态标准，并被认为是生态与节能建筑实践和管理的领跑者。特别是在联建住宅发起的德国南部，这些项目早在生态风潮之前就确立了相关方面的标准。多数联建住宅项目使用了可再生能源，甚至少数项目只使用可再生能源。值得赞许的是，柏林联建住宅项目在早期远在它们需要达到节能生态标准之前就整合了新技术与新标准，将生态品质与建筑品质完美地结合，实现了一体化的设计。

在传统房地产项目中，对于建材与建构的探索是比较缓慢与保守的，这一方面是由于开发商为了追求利益最大化，倾向选用成熟的、性价比高的材料与结构方式，另一方面则是由于政府相关部门在规章制度上的严苛导致开发商不敢轻易越界尝试。而在许多联建住宅项目中，由于项目本身就是由建筑师等专业人员所组织发起的，他们对于建筑的品质与创新有着较高的理想，会想方设法地去实践新的可持续的材料与构造，也会与相关部门进行长期的协商，并最终获得项目的通过。

在能源资源的可持续上，传统房地产开发项目也不乏有许多这方面的探索与尝试，并且也取得了较好的成果，但尝试仅仅局限于项目本身。反观联建住宅项目，不仅每栋

3.3.3 Summary

Germany has been treated as the leading country that practices Eco-energy-saving architecture for a long time. Especially in the southern Germany where the German Co-housing projects derived, relevant standards were established long before the Eco-trend. Most Co-housing projects utilize renewable energy resource and even some only adopt renewable energy resource. Combined with stricter standards and new technology, many Co-housing projects have integrated the Eco-quality and Architecture quality perfectly.

In traditional real estate projects, the exploration on the materials,structure and construction methods are relatively slow and conservative. On one hand it is due to the cost consideration.They tend to adopt mature and secure materials and structure and don't want to pay for the failure of attempts; on the other hand it is due to the strict requirements from the relevant department. However, in Co-housing projects, since many of them are initiated by architects who have a higher demand on the quality and creativity of the buildings, they cherish every chance to test

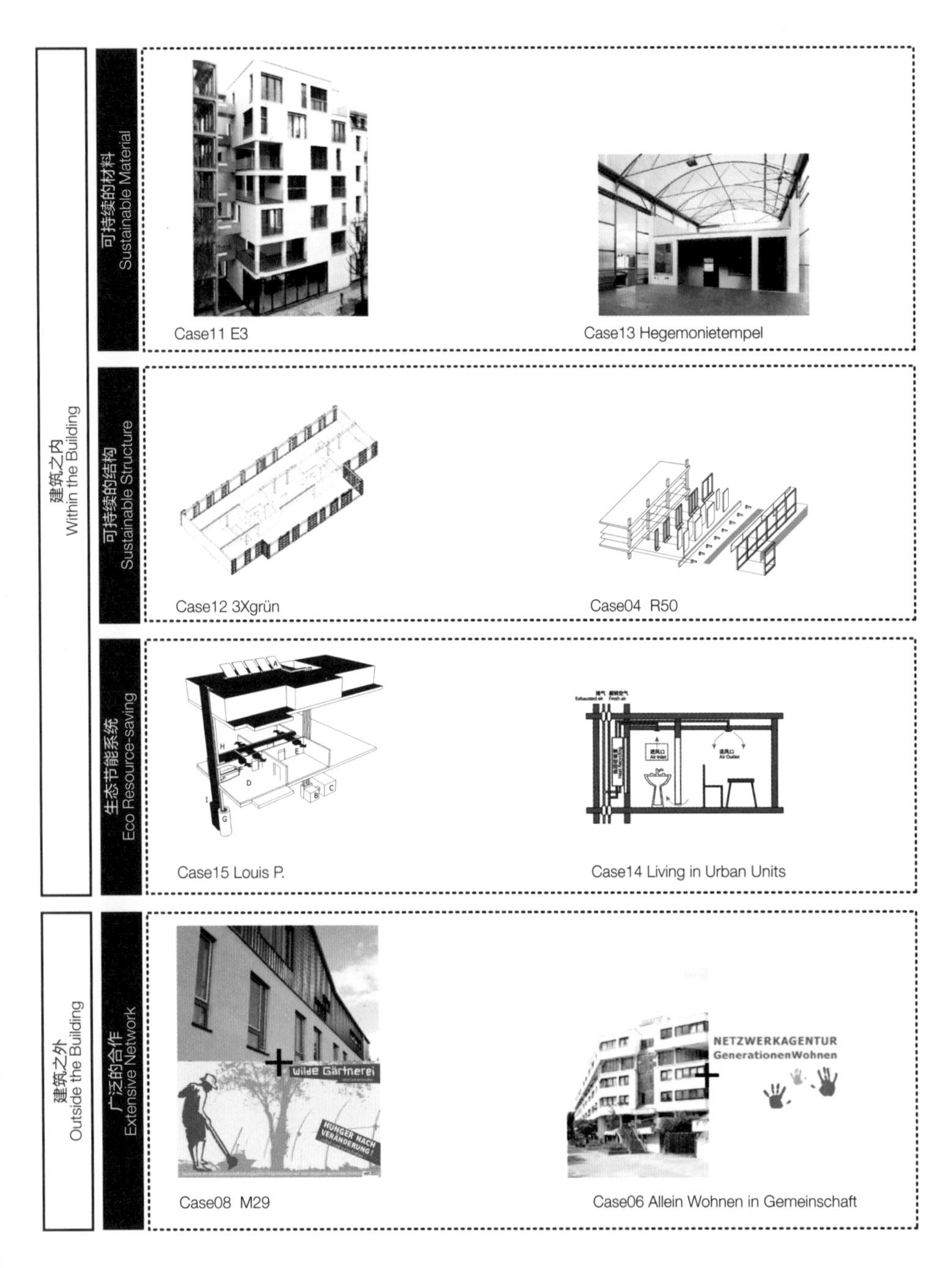

Fig.3.54可持续性的实现策略 Strategy to realize sustainability

建筑都达到了较高的节能生态标准，同时通过广泛的联系，与其他项目及社会组织合作，共享资源能源，提高其利用效率，使能源资源不仅在项目内同时在更大的社会网络中实现可持续化。（Fig.3.53）

通过对已有案例的归纳总结，在实现可持续方面，联建住宅项目的尝试是分两部分（Fig.3.54）进行的，一是建筑之内，一是建筑之外。在建筑之内，主要分为三方面：

（1）**建筑材料**：使用可持续材料，既指生态可持续性环保材料如木材等，又指经济适用性材料如温室材料等，实现经济以及使用方面的可持续性。

（2）**建筑结构**：使用与环保材料性能相适应的结构形式或是利于建筑二次利用的模块化结构，实现使用人群、使用功能的可持续性。

（3）**生态节能系统**：按照相关规划标准，将生态节能系统整合至建筑设计之中，如被动式住宅（Passive House），三升宅（Three-Liter-House）[1] 等，实现能源利用的可持续性。

在建筑之外的部分，则主要通过与社会上的一些相关组织建立合作关系，共享资源、能源以及相关经验。

1　三升宅指房屋每平方米单位面积每年消耗 3 升燃料用于供暖。比较有名的案例是德国巴斯夫“三升宅”。

possibilities of alternative materials and structures. After long time negotiation with the relevant department, in most situations the architects will persuade them with persuasive project schemes.

On the aspect of sustainable energy resources, traditional real estate projects have also made lots of research and achieved positive results, but usually the research scope is restricted within the project itself. In Co-housing projects, not only each building has achieved a high Eco standard but also through extensive networks they share and save the resources and energy in a broader scope.(Fig.3.53)

Based on the selected cases, the attempt to realize sustainability in Co-housing projects can be summarized into two parts (Fig3.54):

One is within the building and the other is outside the building. The former part has three aspects:

(1) **Sustainable Material**: sustainable materials not only contains Eco-sustainable materials like wood but also contains economically sustainable materials. That using the sustainable material, realizes the sustainability on economy and utilization.

(2) **Sustainable Structure**:that using the corresponding structure forms or modular structures that are easy for re-utilization realizes the sustainability on user-groups and functions.

(3) **Eco Resource-saving System**: according to relevant standards that integrating Eco-resource-saving systems into architecture design (passive house, 3-liter-house [1]) realizes the sustainability on resource utilization.

As to the latter part, Co-housing projects establish extensive cooperation with various social organizations to share resources and relevant experience.

1 3-liter-house only consumes 3-liter fuel per sq. for heating each year

传统住宅
被动接受购买价格
Traditional Housing
accept the sale price

联建住宅
主动控制成本
Co-Housing
control the cost forwardly

传统住宅
与社会福利设施无关联
Traditional Housing
have nothing to do with public welfare

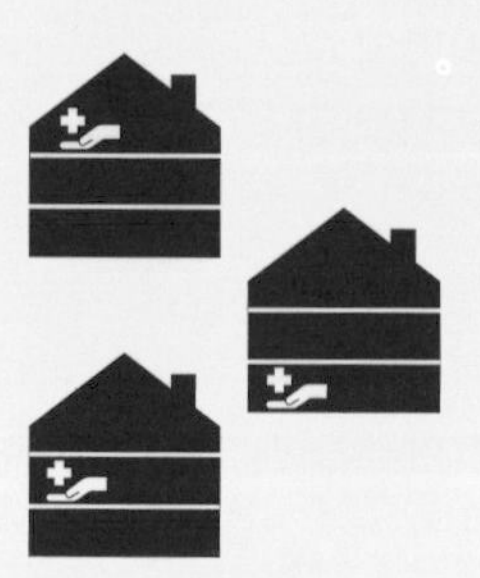

联建住宅
分担一部分社会福利功能
Co-Housing
share responsibility of public welfare

Spontaneity
自发性

need to make their living area and life more efficient and have the sense of community to do something

探索更有效率的生活空间和生活方式，出于归属感和社会责任感为社会出力。

Exploration
探索性

Traditional Housing

accept the sale price passively /
have nothing to do with public welfare

Co-Housing

control the cost forwardly /
share responsibility of public welfare

传统住宅：被动接受购买价格。与社会福利设施无关联
联建住宅：主动控制成本。分担一部分社会福利功能

Fig.3.55自发性与探索性 Spontaneity & Exploration

3.4 自发·实验

联建住宅项目大部分都是自建项目，这使得在建筑上做试验成为可能，而这通常不会在由利益驱动的投资商手上实现。联建住宅项目自发的探索，是出于实验的渴望和爱好。许多专业人士迫切希望探索建筑在规划、组织和技术等方面新的潜力（Fig.3.55）。

3.4.1 自发关注经济成本：项目在控制成本方面的探索

联建项目所面临的重要问题之一即是如何控制成本的问题。不同于面对传统开发商项目被动接受其出售价格的局面，联建团体成员可以通过合理布置使用空间、采用性价比高的建材等一系列手段来主动控制成本，在质量与成本之间获取最佳的平衡点。

3.4 Spontaneity & Exploration

Most Co-housing projects are bottom-up initiated spontaneously and self-made "*Selfmade projects make experiments in building possible, which would normally not be realized by profit-oriented investors....The spectrum covers: the trial of a (super) low-cost building at a high architectural quality, to the development and utilization of new technologies and processes, to the development of participative projects where the user has great freedom and flexibility but also responsibility for the outcome. Alternative solutions are being tested and future-oriented models created, from which we can learn and others can profit in the future.*[1]"(Ring,2013) In this section, the author will discuss the spontaneous exploration on two aspects: Cost Control and Care for the Vulnerable Groups(Fig.3.55).

3.4.1 Focus on Economy spontaneously: Exploration on Cost Control

In the process of practice the most realistic issue Co-housing groups are confronted with is the Cost Control. Unlike people accept the sale price offered by the traditional real estate housing market passively, Co-housing groups can control the cost on their own initiative. By organizing the living space reasonably, adopting cost effective materials & structures and other effective ways, Co-housing projects are trying to find the balance between quality and cost.

1 Source: Ring, K. (2013) Selfmade City Berlin: Stadtgestaltung und Wohnprojekte in Eigeninititie, Berlin: JOVIS: 45

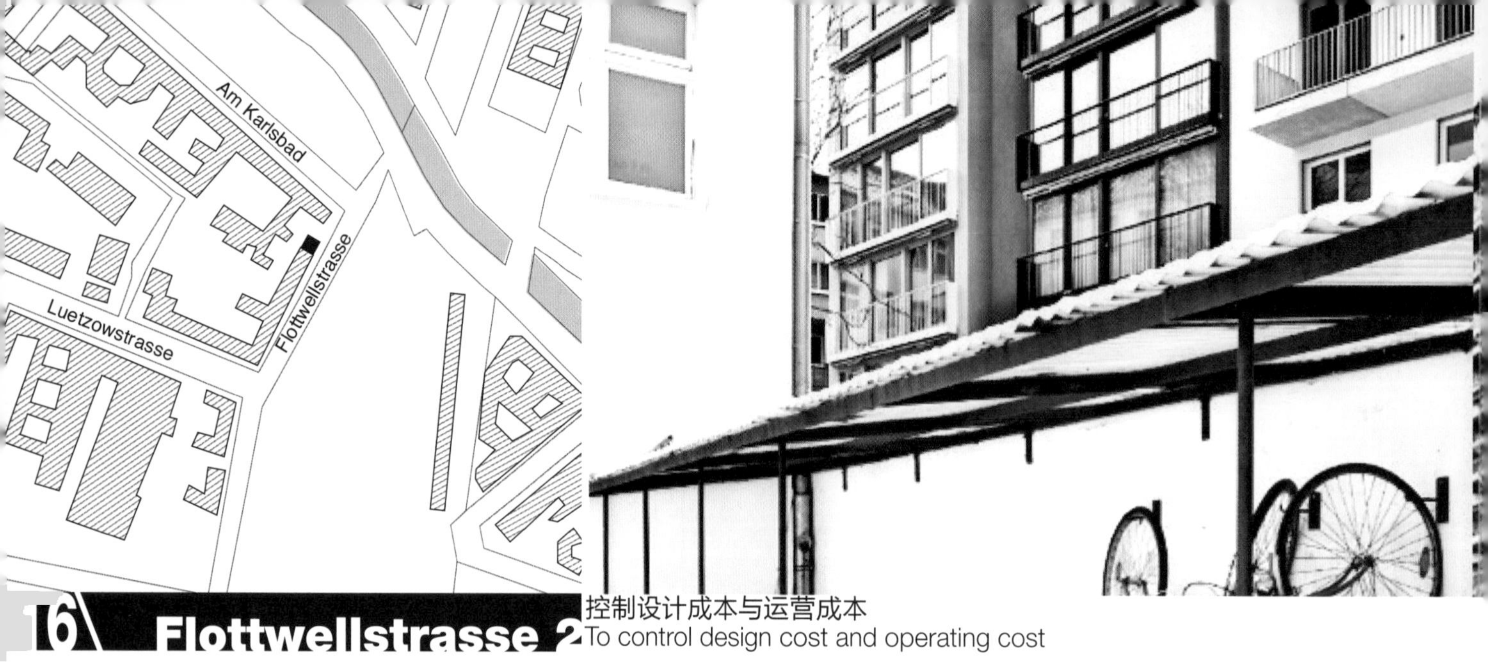

16 Flottwellstrasse 2

控制设计成本与运营成本
To control design cost and operating cost

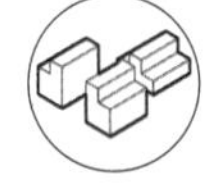

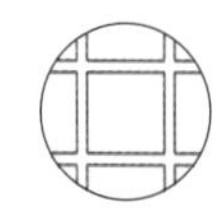

项目概况 Project Profile

项目名称 Project Name：	Flottwellstrasse 2
项目地址 Project Address：	Flottwellstrasse 2
建筑设计师 Architect：	HEIDE & VON BECKERATH ARCHITEKTEN
基地面积 Site Area：	418 m^2
建筑面积 Total Area：	1483 m^2
项目成立时间 Founding Time：	2007
项目状态 Project Status：	完工 finished

Fig.3.56项目区位及实景 Project location & scene

Flottwellstrasse2 号联建住宅项目靠近波茨坦广场(Fig.3.56)，夹在一个居民收入低下的区域和一段前国家铁路基地之间。项目目标是为住宅在经济、生态、社会和文化方面树立新的标准。在控制项目成本方面，大多数项目的惯常做法主要是从设计成本、建设成本和运营成本三个方面进行考虑，本项目则主要是针对设计成本和运营成本进行控制。

建筑由两栋 15m 进深的窄条单元所构成，并在外立面上用颜色加以区分。在设计方面，最关键的是实现设计成本经济性以及居住舒适性之间的平衡。在本案，并没有针对各户进行单独设计而是通过设置标准单元，然后以对其进行扩展的方式来满足不同需求。标准的户型面积为 90m^2，且内部错半层。由于地块限制，建筑平面呈东西向窄条，通过错半层的方式使得阳光在一天中至少有两次能够穿透整个进深，从而提升住户使用的舒适性 (Fig.3.57) 。高差从靠近中间的位置开始，顺着高差沿一侧布置厨房浴室以及独立的卫生间。其他户型以这种户型为基础进行组合，总共有 8 种户型，面积从 32m^2 至 273m^2 不等，适应性强。

在项目完成后的几年里，超过 19000m^2 的居住区将在此建设，铁路区域也将被改造成三角洲公园。因此，建筑底

Project Flottwellstrasse2 is located between a low-income residential area and railway area which is close to Potsdamer Platz(Fig.3.56). This project tries to establish new economical,ecological, cultural and social standards on housing. In terms of Cost Control, most projects always start from three aspects: design cost, construction cost and operating cost. This project then started from the former two aspects: design cost and construction cost.

This building consists of two slim parts which are 15m deep and distinguished from each other by the façade painting. In order to achieve the balance between cost efficiency and living quality, the designers didn't design for each party but a standardized unit based on which lots of unit transformation can be achieved . The standardized unit is duplex with an area of 90m^2. Due to the site limit, the building plan has to be arranged east to west. By the half-level difference, the sunlight can go through the whole unit twice a day thus improving the living comfortability(Fig.3.57) Based on the standardized unit it can transformed into 8 types of units with an area from 32m^2 to 273m^2 adapting to different living demands.

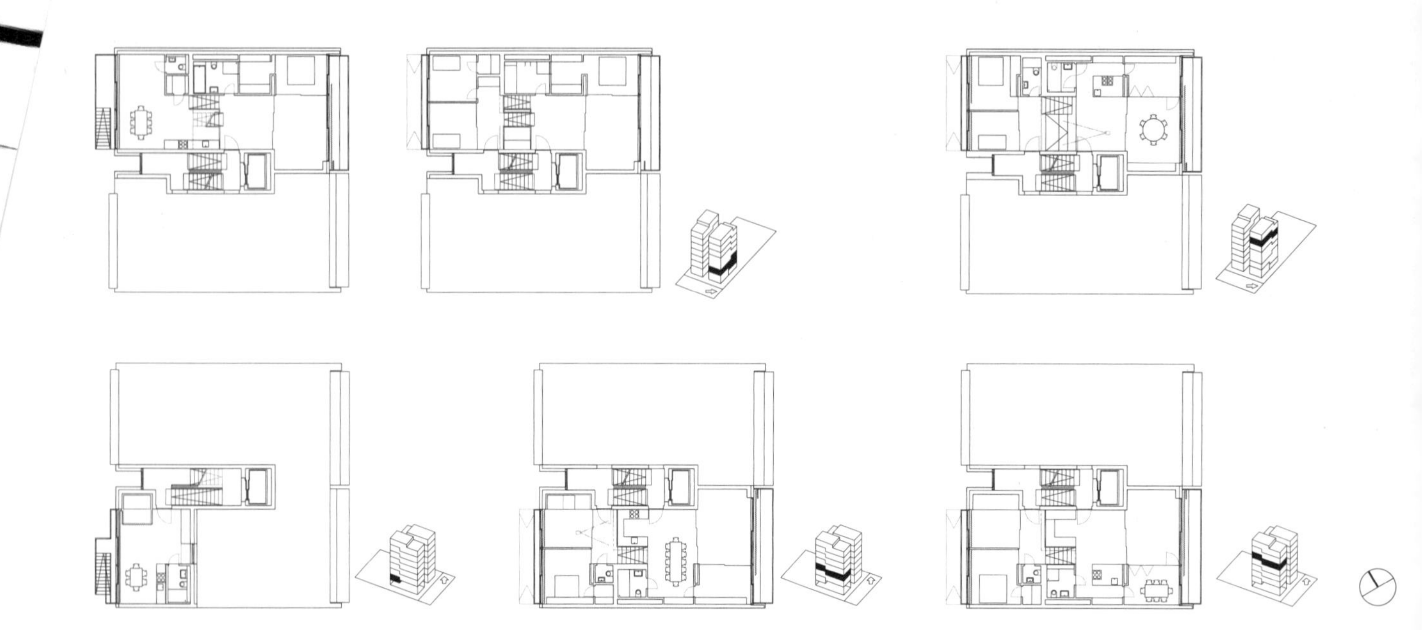

Fig.3.57基于标准模块的个性设计 Individualized design based on standardized module

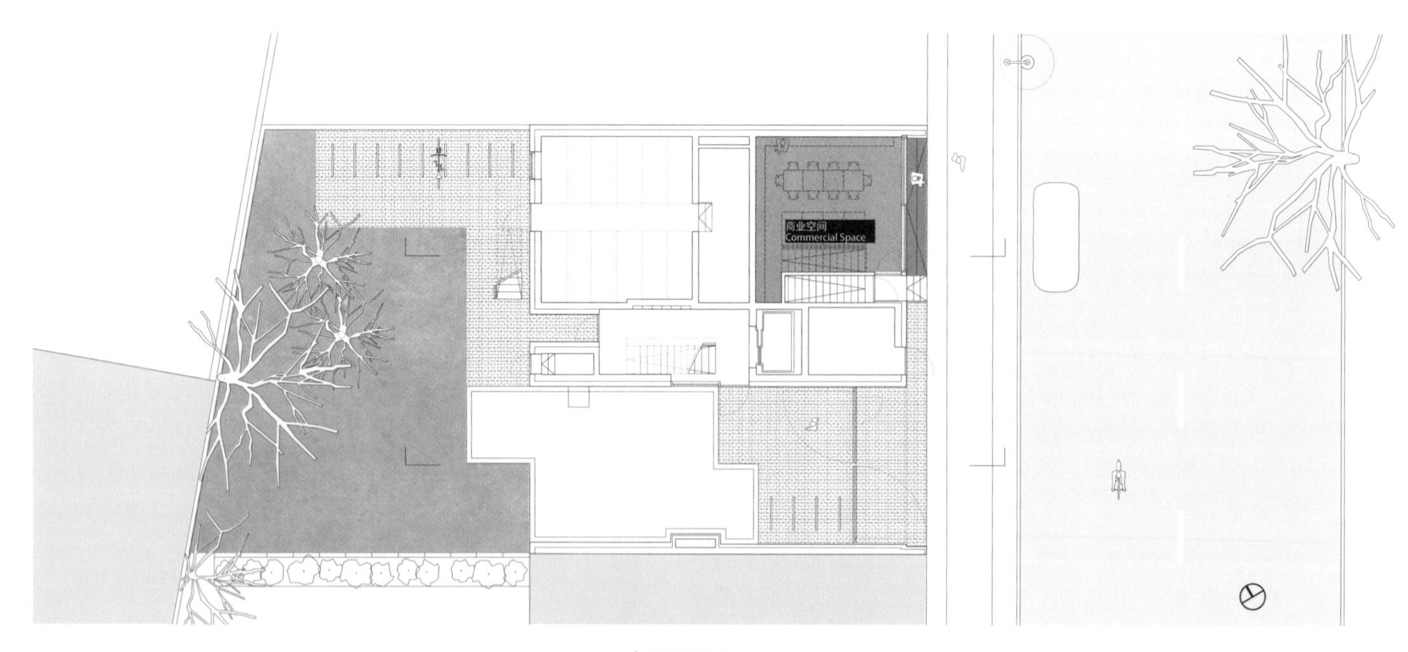

Fig.3.58底层平面图 Ground floor plan

层空间对于项目发起者十分重要，所以他们设置了部分的商业空间用以出租（Fig.3.58），这一方面因为他们意识到自己有机会也有责任为该区域的进一步发展设立标准并且给予其发展动力，另一方面也能在一定程度上缓解来自运营成本的压力。

They are informed that in the following years, over 19000m^2 living community will be built, and the railway area will transformed into a public park. Hence the ground floor space of the building is very important for the initiators and they spare part of the ground floor as commercial area for rent (Fig.3.58). On one hand the group is aware of their responsibility of setting standard for the development of this area and on the other hand it can reduce the operating cost to some extent.

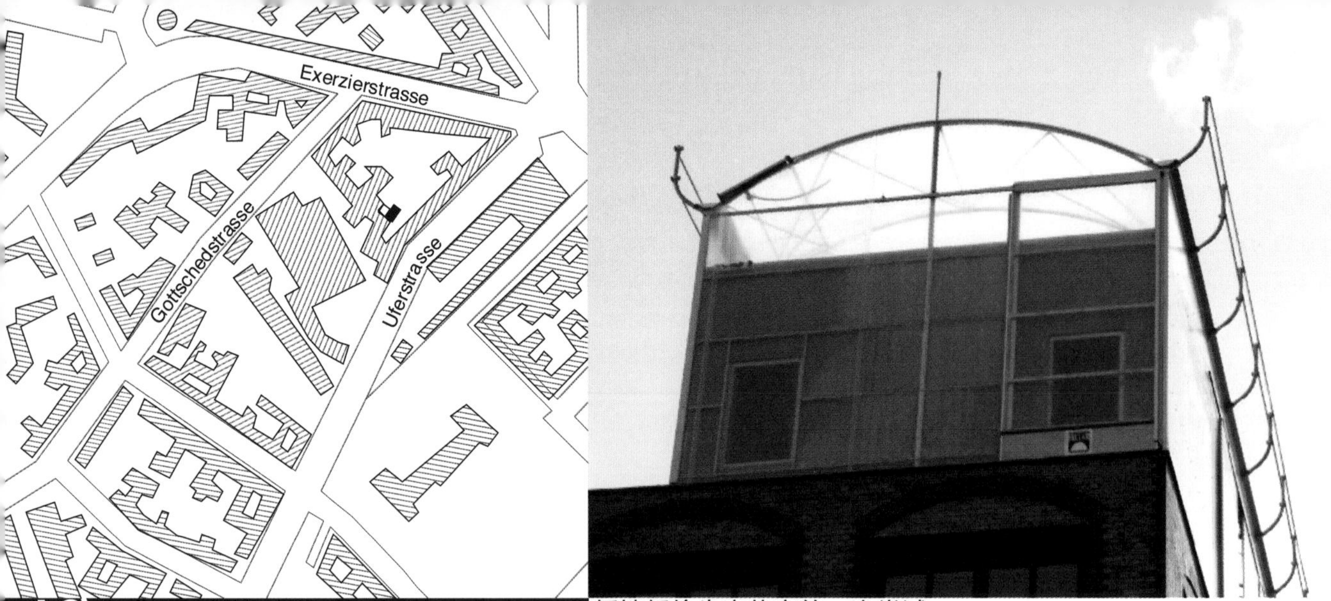

13\Hegemonietempel

低技低价生态住宅的一次尝试
A Low-cost and Low-tech Housing Attempt

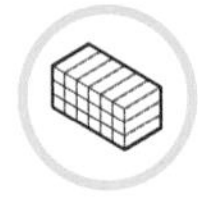
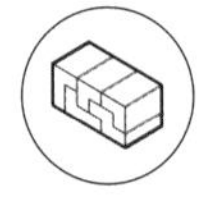
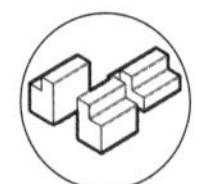

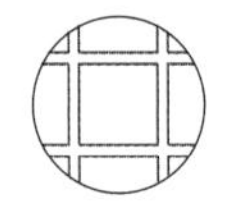

项目概况 Project Profile

项目名称 Project Name：	Hegemonietempel
项目地址 Project Address：	Uferstrasse 6
建筑设计师 Architect：	CHRISTOF MAYER, BUERO FUER ARCHITEKTUR UND STAEDTEBAU
基地面积 Site Area：	283 m^2
建筑面积 Total Area：	134 m^2
项目成立时间 Founding Time：	2009
项目状态 Project Status：	完工 finished

Fig.3.59项目区位及实景 Project location & scene

这座温室住宅正如它看起来一样，像是随时都可买到的半成品，稍事加工就能使用，但是这正是建筑师设计的出发点——设计建造一座低技低造价且生态环保的住宅。项目成员配合建筑师想尽各种办法尽可能地降低每月的租金，最后该项目建筑成本为 70000 欧，每月只需缴纳 375 欧用于用地租金（其中，225 欧用于获得可供住宅落地的 150m^2 屋顶面积；150 欧用于相关设施费用）。

这样的价格已是十分低廉，并且它的使用体验也是令人满意的。生活方式、居住空间以及住宅的使用，可以随着季节变化而变化。在冬天，砖砌房间内保持恒温，而在砖砌房之间的中心开放区域，则通过壁炉取暖。而在夏天这一区域则通过遮阳百叶遮挡过强的日光，避免房间被加热，同时屋顶上的气窗能将室内热气排出。室内很多家具都安上了滑轮，能够自由分隔空间，这对于两位身为艺术家的住户来说再合适不过 (Fig.3.59)。

This Conservatory-like project seems like a half made product which can buy and transform anywhere easily. That is exactly the starting point of the design: to build a low-cost and low-tech Eco-housing. The construction cost is 70000 euro and the monthly operating cost is only 375 euro (225 euro for the 150m^2 roof area's rent and 150 euro for the living). It is quite inexpensive with good use experience.

Living style, living space and the functions change with the seasons. In winter the two brick rooms are heated and in the free-zone between them there is a fireplace to keep the users warm while in summer the free-zone is shaded by the shutters on the roof to block the too much sunlight and the scuttles help exhaust the heated inner air and keep the inner temperature cool. Furniture is equipped with wheels and is used to subdivide the free-zone. The two artists living there are quite satisfied with their "work"(Fig.3.59).

11\ Esmarchstrasse 3

控制建造成本与运营成本
To control construction cost and operating cost

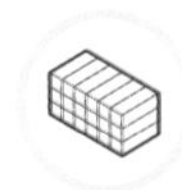  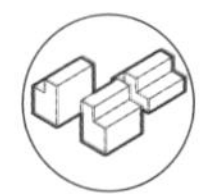

 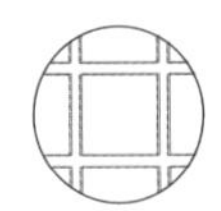 

项目概况 Project Profile

项目名称 Project Name：	Esmarchstrasse 3
项目地址 Project Address：	Esmarchstrasse 3
建筑设计师 Architect：	KADEN-KLINGBEIL ARCHITEKTEN
基地面积 Site Area：	N/A
建筑面积 Total Area：	1270 m^2
项目成立时间 Founding Time：	2006
项目状态 Project Status：	完工 finished

Fig.3.60项目区位及实景 Project location & scene

E3 项目在之前已经介绍过，“城市木构”是这栋建筑的标签。这样一个看上去中规中矩的建筑却是经过了深思熟虑才得以形成。事实表明一方面传统开发过程缺乏实验性，看重短期收益，另一方面建筑界不愿轻易尝试利用新资源。通常情况下即便是刚毕业的建筑系学生也没有在设计中形成能源优化的意识，可持续发展的问题一般是作为建筑的矛盾点进行处理的。E3 项目成员费尽心力寻找经验丰富的事务所帮助他们实现想法。自 2006 年以来，该联建住宅团体走访了多个建筑设计事务所，有一些事务所列举了这样做的一系列弊端：“……长期复杂的审批过程，高消防要求，可能要求安装自动喷水灭火系统或是用金属板密封墙体，高额的建筑保险，昂贵的建筑造价。虽然是木结构，但是你将看不到木材”[1]。最后 Kaden Klingbeil 事务所应允了这一冒险的要求，签订了该项目的合同。

“只有在建造中使用木材才能拯救我们的森林”这是洛桑现代木构建造专家 Julius Natterer 教授的口号，正是他负责了该项目的结构。这并不是仅仅因为木材自然和舒适的属性（这些其实可以通过铺设木质地板来达到），而是因为对于原材料的使用更加有益于生态平衡。更多树木的生长能够吸收更多的二氧化碳，尽管短暂的建造周期成本很贵，但

1　译自 Kleinein, D. (2008) ‘Esmarchstraße 3: Kritische Verkapselung’, Bauwelt, 18/04, 22.

Facts have proved that on one hand traditional real estate projects lack experimental ideas and value short-term gains;on the other hand in the traditional architectural practice architects do not intend to try new resources. In the general situations even the newly graduate students majoring in architecture have no awareness of energy utilization optimization. Sustainable issues are handled as the conflict to design. Based on different intention, Co-housing groups like E3 want to integrate the sustainable ideas into the design forwardly. Cherishing the wish to build a multi-storey wooden structure housing in the inner-city, this Co-housing group visited several architecture offices since 2006. Some reminded them that it would bring a series of obstacles:“*long-term complexed approval process, higher fire-protection requirements with the possibilities of installing automatic sprinkling system or walls sealed by metal pallet, higher architecture insurance and higher construction cost. Besides it is hardly to see the texture of wood from outside* [1]”. Finally KADEN-KLINGBEIL ARCHITEKTEN accepted this project and started to design for the group.

1 Translated by the author from source: Kleinein, D. (2008) ‘Esmarchstraße 3: Kritische Verkapselung’, Bauwelt, 18/04, 22.

这种方式仍然可行。关于木材与石材成本比较的研究由于客户不同其差异性很大。但是如果能在低能耗住宅（例如每年每平方米的能耗为 40kW，即满足德国复兴信贷银行节能标准 40[2]）中使用木材，那么混凝土将失去其价格优势，因为木材能够使得墙变薄，进而只需要较薄的保温层(Fig.3.60)。

该建筑每平方米造价 2250 欧，属于正常范围，低于市场价格每平方米 2500 欧。但是每年的供暖费用只有 500 欧，这对于每户 $140m^2$ 的公寓来说是长期的价格优势。

2 KfW-40：每年的能耗需求符合 EnEV 标准建筑的 40%

"Only using the timer as the construction material can save our forests" is the motto of Professor Julius Natterer [2], who is in charge of the architectural structure of the project. This is not only because of the natural and comfortable attributes of wood which can achieve simply by laying wood parquets,but also because using the raw materials are more beneficial to sustain the Eco balance than other materials. More trees means more CO2 consumption. Even though the construction cost is relatively higher than usual, it is still feasible. If they are used in the low energy consuming housing (for example 40 kWh/m^2 per year), which meets the requirements of KfW-40 (annual energy consumption reaches 40% of EnEV[3] standard architecture), then concrete will lose its price advantages, because timber can make walls thinner which leads to thinner insulating layers(Fig.3.60).

The construction cost of project E3 is 2250 euro which is within the normal range, but the yearly heating fee is only 500 euro which is long-term economical for an apartment of $140m^2$.

2 Julius Natterer, expert on modern wooden structure construction in Lausanne, Switzerland

3 EnEV: The Energieeinsparverordnung (EnEV), or Energy Conservation Regulations, is Germany's energy effi-ciency building code. One of the most stringent codes in the world, the EnEV sets standards for insulation, fenestration, envelope, and HVAC. The code passed originally in 2002, and meets requirements for the EU EPBD and was revised in 2009. (Source: http://energycodesocean.org/code-information/germany-energy-conservation-regulations-buildings-enev-2009)

3.4.2 自发关注社会焦点：项目在人文关怀方面的探索

Stiftung Trias 基金会成员 Rolf Novy-Huy 曾提到："我们认为联建住宅主要的挑战来自于弄清城市发展中的益处。提及城市发展，对于柏林来说这是一个好机会，因这能够使个人资本生效。以个人理想主义为动力的联建住宅项目积极地展示它们的社会效应，挑战不仅在于要关注代际、老人和妇女问题，也要关心残障人士……"[1]。通过帮助这些相对弱势的群体结成联建住宅团体，从而更好地帮助其改善生活品质。

1　作者译自 Ring, K. (2013) Selfmade City Berlin: Stadtgestaltung und Wohnprojekte in Eigeninititie, Berlin: JOVIS.

3.4.2 Focus on Social Issues spontaneously: Exploration on Care for Vulnerable Groups

"From our point of view, the main challenge for Co-housing projects is to develop and clarify the benefits for urban development. In terms of urban development, this is an opportunity for Berlin to bring private capital into effect. The collectively organized housing projects characterized by idealist motives, are striving to demonstrate - as is the case everywhere in Germany - their social benefit. The challenge is to take into consideration not just the issue of generations, the elderly, and women, but also increasingly those in care and people with disabilities.[1]" (Novy-Huy[2],2013)

1 Source: Ring, K. (2013) Selfmade City Berlin: Stadtgestaltung und Wohnprojekte in Eigeninititie, Berlin: JOVIS: 203

2 Rolf Novy-Huy, director of Stiftung Trias

17\ Südwestsonne

为疾病晚期患者提供良好的生活环境
To Provide Good Living Space for Incurably ill Patients

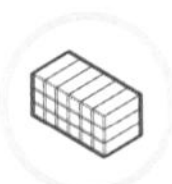
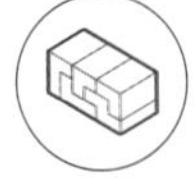
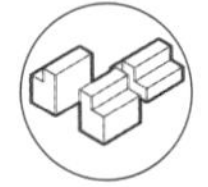

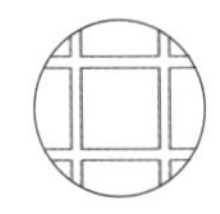

项目概况 Project Profile

项目名称 Project Name：	Südwestsonne
项目地址 Project Address：	Scharnweberstrasse 45
建筑设计师 Architect：	CHRISTOF MAYER, BUERO FUER ARCHITEKTUR UND STAEDTEBAU
基地面积 Site Area：	516 m^2
建筑面积 Total Area：	1400 m^2
项目成立时间 Founding Time：	2007
项目状态 Project Status：	完工 finished

Fig.3.61项目区位及实景 Project location & scene

大部分联建住宅项目都在自己力所能及的范围内回馈城市以及社会，本案即是一例。项目位于靠近法兰克福大街的一条安静的街道上，是一栋出租房和产权房混合的住宅(Fig.3.61)。建筑靠近庭院一侧朝向西南面，将来通过阳台可以看到不远处即将建起的佛寺，庭院旁的共享空间由住户共同使用。

建筑采用预应力混凝土框架结构，这为平面设计提供了极大的灵活性。生态环保理念不仅表现在重视建筑外立面与内饰生态建材的使用，同时也体现在对于替代性和资源节约型能源的使用。

住宅项目的底层在面向街道处设置了项目团体的办公室，而二层至五层为公寓，部分设计成了出租型公寓以便项目团体定期重组，以产生新的混合可能性。

项目一层有五间小公寓、一间客房以及一个共享的起居室和厨房，这一层以低价出租给项目团体成员之一的 NiWo e.V. 协会 (独居终老住房协会)。该社会团体致力于为那些罹患末期疾病的人们提供良好的生活环境，让他们生活在社区之中而非被隔离，能够自己决断生活而非依赖他人，同时获得租金稳定的住所。此外该协会的工作人员还为本项目以及周边邻里的患者提供紧急救助以及善终服务。

Project Südwestsonne is located in a quite street (Fig.3.61) . The southwest side of the building faces the shared courtyard and from the balcony in the future a Buddhist temple will be seen. The building adopts pre-stressed concrete frame structure which allows flexible plans. Ecologically environmental-friendly concept not only reflects on the Eco-construction material utilization but also reflects on the utilization of alternative and resource-saving energy.

On the ground floor there is an office for the Co-housing group while on the second to fifth floors are the apartments with the combination of rental type and private type in order to create new mixed possibilities when the group members change. The first floor is spared for the NiWo e.V. [1] who is also the member of the Co-housing group.This association is devoted to providing those people who suffer from incurable illness with good living environment. In this floor there are five apartments, a guestroom, a shared living room and a shared kitchen. NiWo e.V. also provides service like first aid and Hospice for the neighborhood.

1 Netzwerk Integriertes Wohnen Verein fuer selbstbestimmetes Wohnen und Leben bis zuletzt (Network of Inte-grated Housing Association for independent life and living up to the last) Website: http://www.niwo-berlin.de/

18 Müggelhof Friedrichshain

女子联建住宅项目
A Single Women Co-housing Project

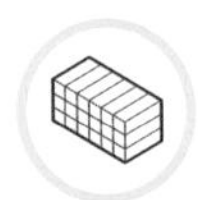
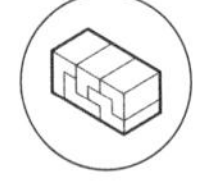
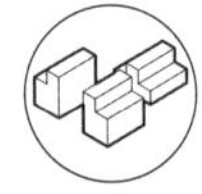

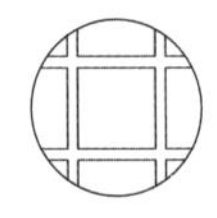

项目概况 Project Profile

项目名称 Project Name： Müggelhof Friedrichshain
项目地址 Project Address： Müggelstrasse 21
建筑设计师 Architect： Stefanie Ruhe
基地面积 Site Area： 1500 m^2
建筑面积 Total Area： 1800 m^2
项目成立时间 Founding Time： 2008
项目状态 Project Status： 完工 finished

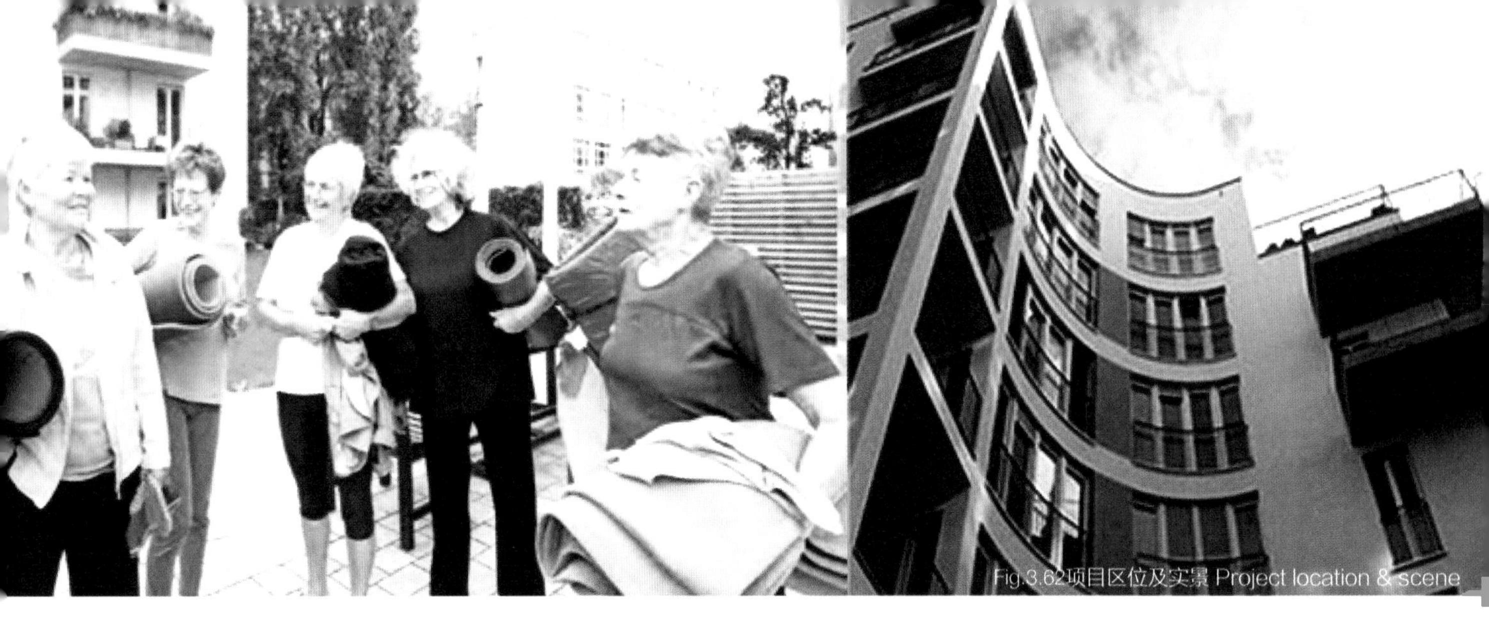
Fig.3.62项目区位及实景 Project location & scene

在柏林有超过六十万女性是独自生活的，这种现象所带来的安全、健康等问题值得关注[1]。而Müggelhof Friedrichshain 项目是基于柏林妇女独立自主传统的联建住宅项目，它向所有年龄、所有宗教信仰以及生活方式的妇女开放，旨在帮助她们塑造多样、公平、安全以及宽容的社区。

项目搭建了一个经验、思想、知识交流平台，居民可以小规模或大规模组织在一起，共同学习，互相帮助，提高生活的品质。

该项目大部分为小户型，供个人居住，还有一些大户型供集体居住。房屋产权归个人所有（部分人将其出租给其他女性）。在社区内有公共厨房、浴室、客房和花园等共享空间，为住户的社区生活提供了必要的场所 (Fig.3.62)。

In Berlin over 600,000 women live independently, and this phenomenon brings a lot of security and health issues which need to be paid attention to. Project Müggelhof Friedrichshain is based on the tradition of women's self independence in Berlin and is open to the women of all ages, religions and living ways aiming at helping them build up a multiple, equal, secure and tolerant community.

The project creates a platform for exchange of experience, ideas and knowledge. The group members can learn together and help each other either in a small group or in a bigger scale to improve the quality of life.

It offers small units for the individual and large units for collective living. The ownership is private and some of them have it rent to other women. There is a communal kitchen, bathrooms, guestrooms, gardens and other shared space for community life(Fig.3.62).

1 来源：www.beginenwerk.de/Mueggelhof_Berlin/index.html

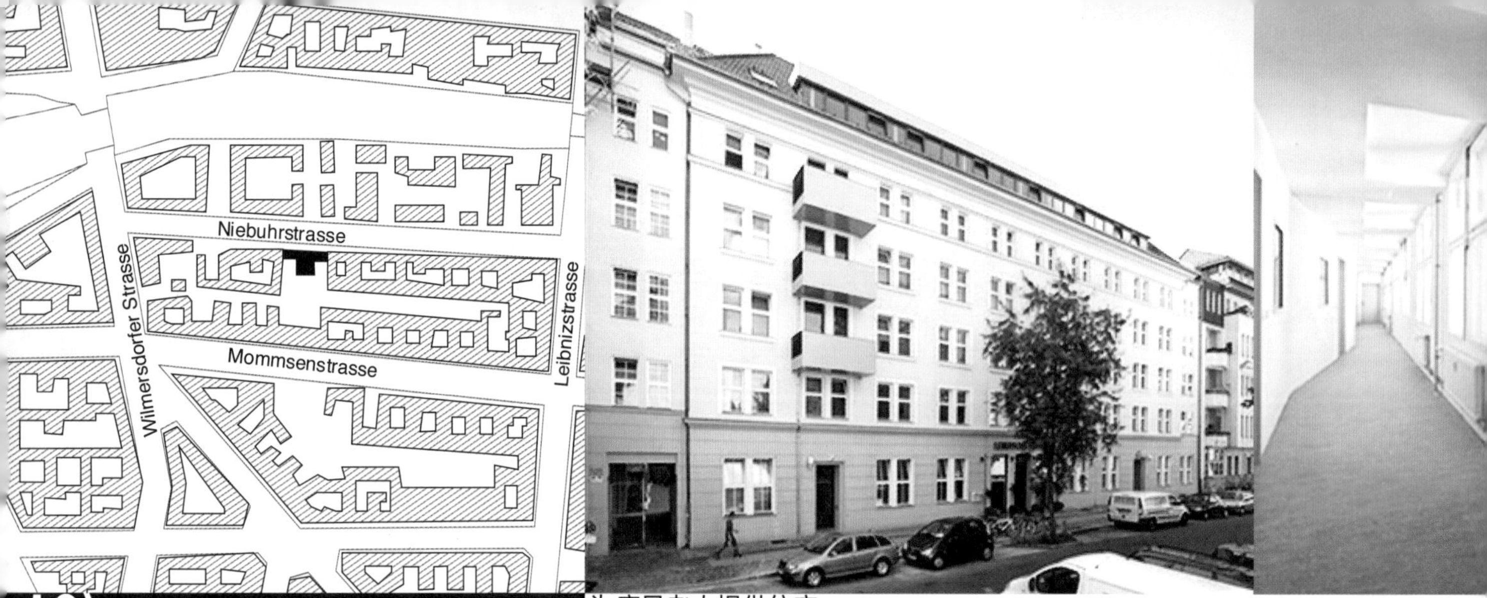

19\ Lebensort Vielfalt

为痴呆老人提供住宿
Accommodation for Old Men suffering Alzheimer's Disease

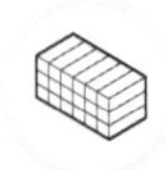
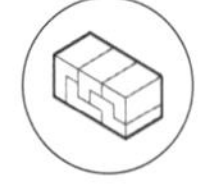
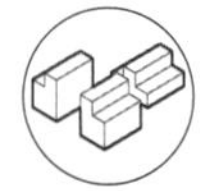

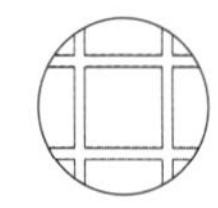

项目概况 Project Profile

项目名称 Project Name：	Lebensort Vielfalt
项目地址 Project Address：	Niebuhrstrasse 59/60
建筑设计师 Architect：	ROEDIG SCHOP ARCHITEKTEN
基地面积 Site Area：	2230 m^2
建筑面积 Total Area：	5098 m^2
项目成立时间 Founding Time：	2007
项目状态 Project Status：	完工 finished

Fig.3.63项目区位及实景 Project location & scene

该案例是一个改造项目，原建筑建于1939年，用于居住和银行办公，而在1959年至2006年间则用作日托中心(Fig.3.63)。建筑的底层和一层经改造后，为老年痴呆症患者提供住所，二层及以上则被改造成为代际居。项目总共拥有25套住房，其中2套是为残疾人提供的，15套考虑了无障碍设计。底层的前银行办公大厅被改造成为了咖啡馆和图书馆，为住户和当地居民提供服务。居民之间相互帮助照看，在一定程度上也缓解了这些患者对于社会及政府的压力。

Project Lebensort Vielfalt is a renovated case. The original building was built in 1939 and used for residence and bank office(Fig.3.63). During 1959 to 2006 it was used as a day-care center. After transformation the ground floor and the first floor are used to accommodate lonely old men suffering Alzheimer's Disease. The second floor and above are transformed as multigeneration housing. The project contains 25 apartments 2 of which are for the handicapped and 15 of which are barrier-free. The lobby of the former bank is transformed as Café and library offering service for the group and the neighborhood. The residents help and support each other, reducing the pressure to the society and the government to some extent.

传统住宅
被动接受购买价格
Traditional Housing
accept the sale price

联建住宅
主动控制成本
Co-Housing
control the cost forwardly

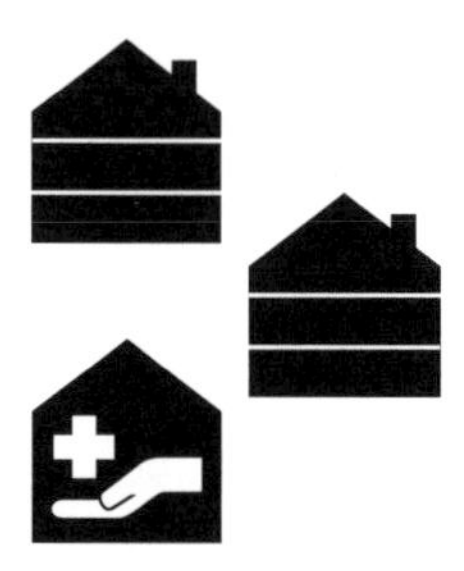

传统住宅
与社会福利设施无关联
Traditional Housing
have nothing to do with public welfare

联建住宅
分担一部分社会福利功能
Co-Housing
share responsibility of public welfare

Fig.3.64联建住宅与传统住宅在自发性与探索性上的区别
Difference between Co-housing and traditional housing on Spontaneity & Exploration

3.4.3 小结

联建住宅的自发性，使得其团体成员不再被动接受不尽如人意的现状，而是主动参与到改造现状的实践中，探索各种可能性。在实践过程中，并不是没有限制，例如成本问题，就是联建团体必须考虑的一个重要问题（Fig.3.64）。通过对已有案例的分析总结，在这一问题上主要通过以下几个方面来控制（Fig.3.65）：

（1）**控制设计成本**：

以灵活的平面、结构以及可供扩展的标准单元通过分隔、组合等手段来适应不同人群的居住需求，降低项目前期所带来的设计成本以及项目运营过程中使用人群更替所带来的改造费用。

（2）**控制建造成本**：

采用可持续的建材与建构进行项目的建造，使用可持续材料（减少在保证建筑使用舒适性方面的投入，如减少保温层厚度）等；在既有建筑基础上进行改扩建，例如利用一些烂尾工程已经建设的基础，利用废弃建筑的屋顶空间等。

（3）**控制运营成本**：

使用可持续能源、资源以及节能环保系统实现运营过程中能源利用率的提升以及能源消耗的降低；通过与社会组织结成合作关系，共享部分资源及能源；出租预先设计的商业文化空间，减缓日常运营成本压力。

主动控制成本是在传统开发商项目中难以实现的，另外联建项目的自发性使得团体成员往往对项目本身以及所处街区、城市乃至整个社会都肩负责任感，所以他们也会关注一些诸如养老、病患医疗等社会焦点问题，希望能够通过项目本身，尽自己所能（如通过开放部分空间资源，承担社会福利机构的一部分功能），帮助社会弱势群体，让他们参与到项目中来。大部分联建项目都是有理想、有思想的实践项目，通过这些方式尽自己所能一点点改变周边环境。

3.4.3 Summary

The spontaneity of Co-housing means the group members take part in the practice of transforming the reality actively instead of accepting the unpleasant situations passively. In this process they might confront different issues like the cost control(Fig.3.64). Based on the selected cases, there are three major aspects to control the cost (Fig.3.65):

(1) **Design Cost Control**: by adopting flexible plans, structures and extensible standardized units, the designers manage to meet the different clients' demands and lower the design cost in the preliminary phase and the potential transformation cost caused by the change of the users.

(2) **Construction Cost Control**: by adopting sustainable materials and structures to construct the buildings, Co-housing projects manage to control the cost as follows: making use of the attributes of materials, existing buildings' transformation, unfinished buildings' built parts like foundation and basement and the roof space.

(3) **Operating Cost Control**: by adopting sustainable energy, resources and energy-saving systems Co-housing projects manage to promote the energy utilization and lower the energy consumption; by establishing extensive cooperation with various social organizations, Co-housing projects share experiences and resources; by predesigning rental commercial and cultural space, Co-housing projects manage to reduce the operating cost pressure.

Controlling the cost actively can hardly be realized in traditional real estate projects. Besides, Co-housing projects' spontaneity makes the members feel the strong responsibilities not only for the project itself, but for the urban block where they are living in even for the society. That's why they often show their care for the social issues like aged

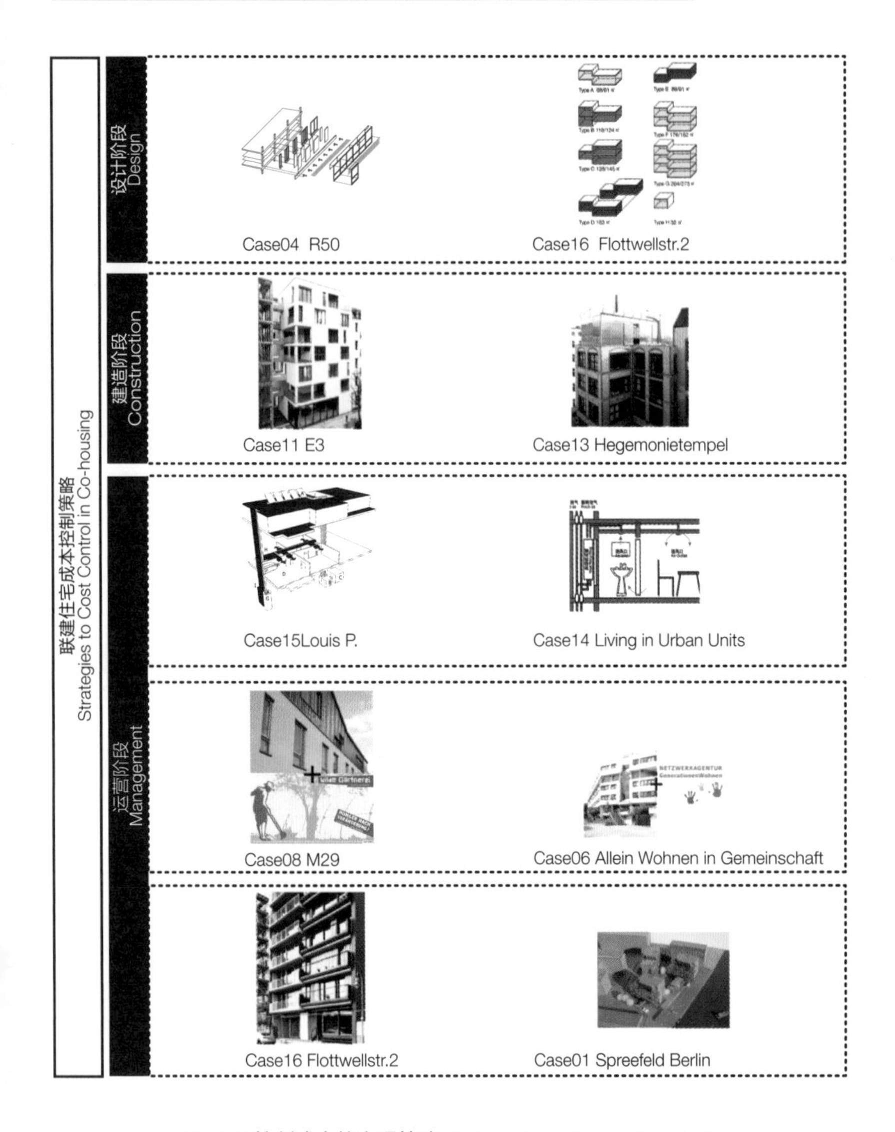

Fig.3.65控制成本的实现策略 Stategy to realize cost control

support and vulnerable groups' problems. By offering shared space back to the city, sharing partial function of social welfare entities and so on, they help and support the vulnerable groups and make them part of the projects. By their own ways, Co-housing projects change the surroundings little by little and they are the practitioners with ambitions.

Conclusion 总结 Part 4

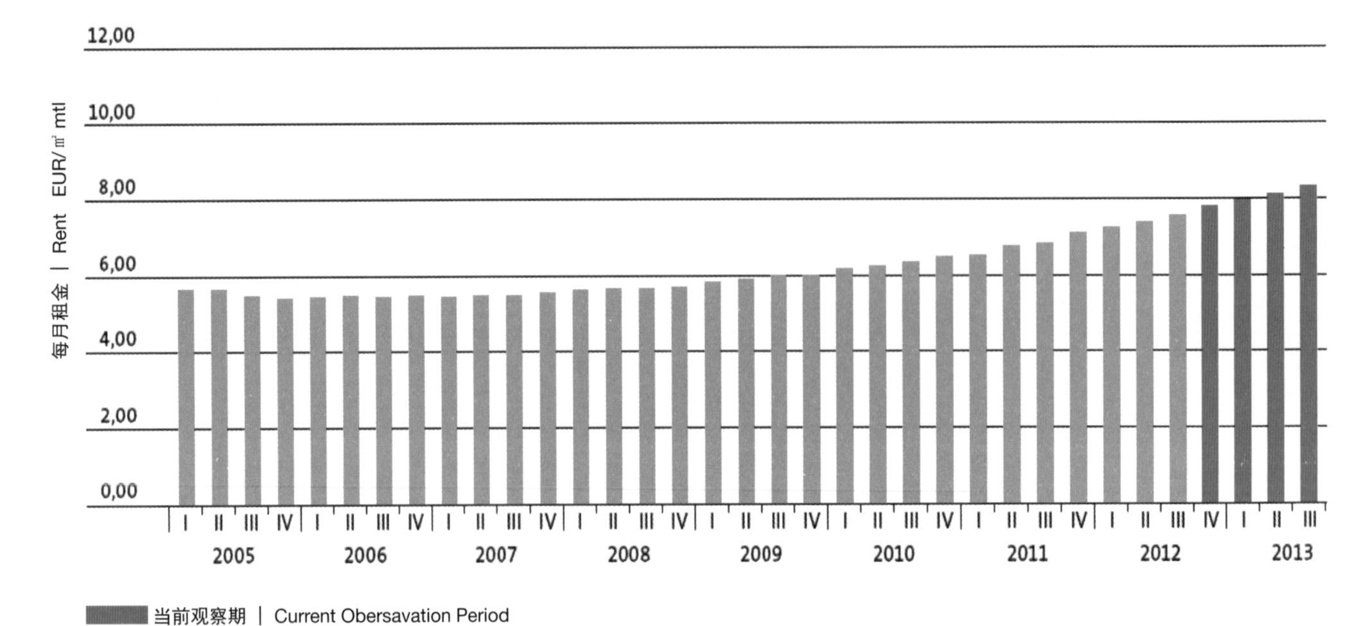

Fig.4.1 2005–2013 年柏林出租房基础租金变化 basic rent changes in Berlin from 2005-2013

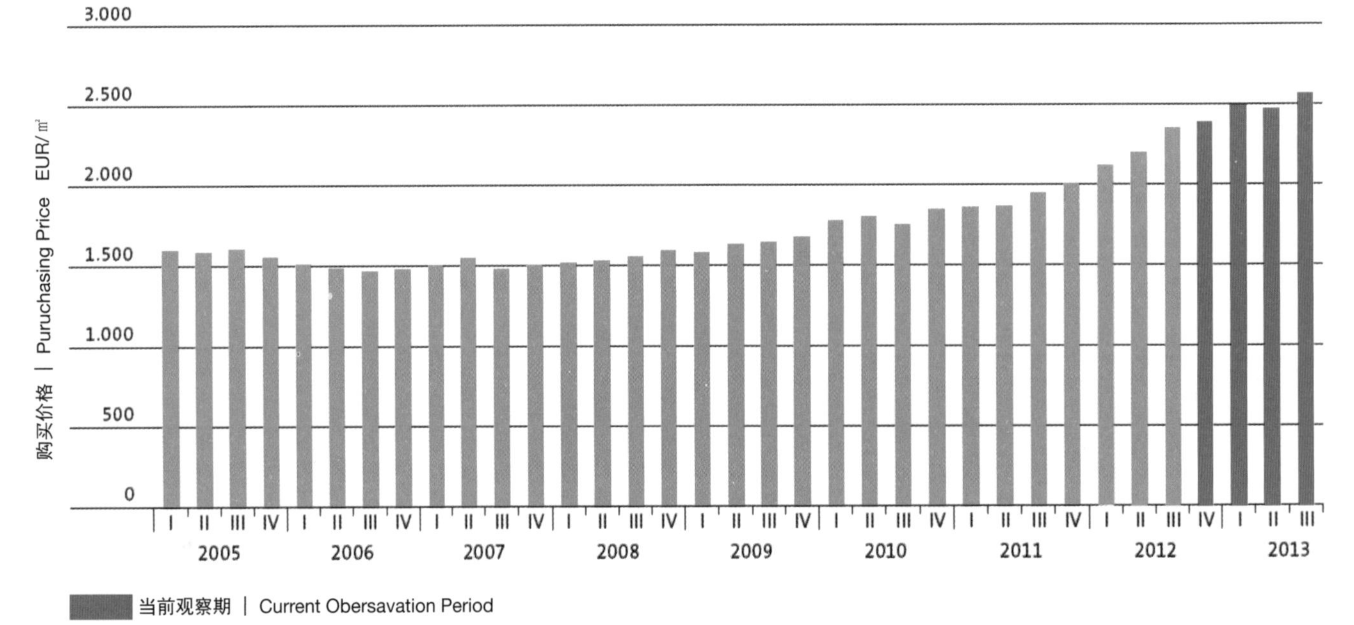

Fig.4.2 2005–2013 年柏林私有住房售价变化 Housing purchasing price changes in Berlin from 2005-2013

4.1 柏林联建住宅的意义

社会、城市、形式与空间以及可持续与生态，是先锋住宅项目的四个本质属性，本节将讨论柏林联建住宅在这四方面的意义。

4.1.1 柏林联建住宅在社会方面的意义

第一，联建住宅项目补充和丰富了柏林既有的住房形式，为中低收入阶层提供了长期稳定并且可负担的住房来源。柏林新建住宅的方式主要有三类，一类是自建房（每户居民可以建造规定大小和数目的住宅而不需要经过相关部门的审批），一类是政府投建的社会福利住房，以及由房地产开发公司新建的住房。联建住宅的出现扩展了自建房的定义，丰富了既有新建住房的方式。另一方面，柏林 86% 的住房为出租房，14% 的住房为私人自住。柏林约 73% 的房屋产权为私人所有，这其中也包含了部分联建住宅，而 16% 的房屋产权在六大市政住宅公司手中，它们也是社会住房的重要合作方，剩下 11% 的房屋产权则在众多的住房合作社手中，它们也是联建住宅项目的重要组成部分 [1]。联建住宅项目已经成为柏林重要的住房形式；从 Fig.4.1 可以看出出租房基础租金一直在小幅上涨，2013 年超过了 8 € /m², 而 GSW [2]2012 年的报告显示在克劳茨伯格、弗里德里希以及米特区等市中心的房租仅一年就上涨了 10%，约为 11.67 € /m²。因而越来越多的人倾向购买产权房以确保长期稳定的生活成本。而根据 Fig.4.2 可以看出私有住房的售价也在小幅上涨，2013 年在 2500 € /m² 附近徘徊。而以联建住宅项目的方式建造的造价大部分介于 2000~2500 € /m² 之间，有些甚至是以 1500 € /m² 左右的造价完成了项目，所以在造价方面的优势以及由此而带来的附加值（高度订制，社区生活等）使得联建住宅项目越来越受到关注。联建住宅项目的初衷之一就是将住房永久地隔离自市场之外，不再充当商品而仅仅是提供居住服务。不管是基于自用的产权房模式，还是出租房模式，联建住宅项目

1 来源：SenStadtUm 2012_Wohnen

2 Gemeinnuetzige Siedlungs- und Wohnungsbaugesellschaft Berlin mbH, 德国房地产公司

4.1 Significance of Co-housing in Berlin

In this section the author tries to follow this method and discusses the significance of Co-housing in Berlin according to these four aspects.

4.1.1 Significance in Social Aspect

Firstly, Co-housing projects has supplemented and enriched the existing housing forms in Berlin and provide middle and low income groups long-term affordable and stable source of housing.In Berlin there used to be three ways of building new housing: one is self-made housing (each household can construct prescribed amount and size of house without approval of relevant department), one is social housing subsidized by the state and another is the real estate housing projects. The emergence of Co-housing has broadened the definition of self-made housing and extended its scale while ways of building new houses have gained. According to a report [1] in 2012, Berlin is a city of tenants. 86% of housing are for rental while 14% are private ownership. About 73% of the housing property in Berlin are hold by the private owners containing parts of Co-housing group members. 16% of the housing property in Berlin are hold by 6 municipal housing companies which are the important partners of social housing. The rest 11% of the housing property in Berlin are hold by many housing cooperatives which are the great components of Co-housing projects. This situation shows that Co-housing has become an important way of building home in Berlin. Fig.4.1shows the continuous rise of the rent in Berlin, and the monthly basic rent (heating fee excluded) in 2013 exceeded 8€/m² while a report of GSW [2] in 2012 showed that rent in Kreuzberg, Friedrichshein and Mitte rose by 10% to 11.67€/m² after one year. That's why more and more people tend to purchase private housing for long-term

1 SenStadtUm 2012_Wohnen

2 Gemeinnuetzige Siedlungs- und Wohnungsbaugesellschaft Berlin mbH

都使得住宅建筑的造价或是租金都长期维持在低于市场价格的范围之内，为中低收入阶层提供稳定的住房。

第二，联建住宅项目为柏林内城发展过程中出现的一系列问题（老龄社会问题、社会医疗福利问题、儿童妇女安全问题等）提供了值得借鉴的解决策略。人口老龄化是欧洲国家和城市面临的严峻现实，预计到 2030 年，柏林平均人口年龄从 2011 年的 42.3 岁上升至 44.2 岁，65~80 岁人口将增加 14%，80 岁以上的人口将增加 80% [3]。如何养老成为重要的问题，养老院模式并不适用于所有人，大多数人仍然希望自己老年能够在熟悉的社区氛围中（临近朋友，亲人）独自生活。联建住宅项目提供了有别于养老院模式的其他解决策略，如老年公寓型养老模式，多代居型社区养老模式等都是利用社区的资源，以及通过社区建立的人际网络为老年人生活提供安全可靠的保障，同时低廉稳定的日常生活成本（包括租金）极大地减轻了老年人的生活压力；另一方面联建住宅项目还为社会上的一些弱势群体如妇女、儿童以及病患提供了较为安全便利的生活环境；同时联建住宅项目还主动地分担社会福利机构的一部分功能，通过低价出租居住空间给相关福利机构以及开放部分共享空间给社区，改善邻里空间的生活品质。

第三，联建住宅项目整合了分散的个人资本、实验性的学术研究成果等社会资源，为拥有想法的先锋个人提供了城市实践的机会和平台，提出了更加优化的解决策略。联建住宅项目通过集合联建团体成员的分散资金大大降低了新建住宅的个人启动资本，使得很多中低收入居民有机会参与到建设自己家园的实践中来；同时联建住宅项目的规模使得小范围的住宅实验实践成为可能，一些在传统开发商项目中无法得以实践的学术研究成果也找到了实践的平台，在一定程度上推进了住宅项目在结构选型、用材等方面的发展。

3 来源：SenStadtUm 2012, Bevoelkerungsprognose fuer Berlin und die Bezirke 2011-2030, Senatsverwaltung fuer Stadtentwicklung und Umwelt [online]. available: http://www.stadtentwicklung.berlin.de/planen/bevoelkerungsprognose/download/bevprog_2011_2030_kurzfassung.pdf [accessed 10 Jan 2014]

stable living cost. According to Fig.4.2 the selling price of private housing also rose and in 2013 it fluctuated around 2500€/m^2 while construction cost of most Co-housing fell between 2000~2500€/m^2 and even some projects finished with 1500€/m^2. The price advantages and its bonus (highly customer-fit and community life) make people turn to Co-housing methods to build homes. One of the original intentions of Co-housing projects is to exclude housing permanently from the speculating markets as commodities. Whether in rental model or private property model Co-housing projects keep the rent or the construction cost below the normal market price to offer people stable housing source.

Secondly, Co-housing projects provide alternative and innovative solutions to the series of issues appearing in the inner city development (aging society, social medical welfare, children and women safety and so on). Aging of population is the severe situation that Europeancountries and cities are facing. According to a relevant report [3] in 2012, the average population age will rise from 42.3 years old in 2011 to 44.2 years old in 2030 and the age group of 65-80 will increase by 14% and the age group above 80 will increase by 80%. That how to support the aged is an important issue and the retirement house model is not suitable for everyone. Most people would like to live independently in the familiar community near their relatives and friends when they are getting old. Co-housing projects offer solutions different from the retirement house model. All these projects make full use of community resources and the social network established through community lives to guarantee a safe and reliable living environment for the elderly meanwhile the affordable and stable living costs relieve the economic pressures

3 SenStadtUm 2012, Bevoelkerungsprognose fuer Berlin und die Bezirke 2011-2030, Senatsverwaltung fuer Stadtentwicklung und Umwelt [online]. available: http://www.stadtentwicklung.berlin.de/planen/bevoelkerungsprognose/download/bevprog_2011_2030_kurzfassung.pdf [accessed 10 Jan 2014]

4.1.2 柏林联建住宅在城市设计方面的意义

第一，联建住宅项目补充和丰富了城市共享空间的内容和形式，为提升区域性认同感和责任感奠定了空间基础。联建住宅项目在一定程度上打破了公有与私有空间的界限，在大尺度的城市层面共享空间与小型社区层面共享空间之间插入了一级共享空间作为二者的过渡；同时联建住宅项目还丰富了城市共享空间的内容和形式，将休憩、娱乐、教育、文化、社区服务及城市实践等方面的功能在这一层面进行整合，室外的半室外的，形式丰富多样，并辐射至邻里，成为凝聚社区积极关系的重要工具。

第二，联建住宅项目将居民、街坊邻里引入城市设计的决策过程，积累公众参与的经验教训，完善参与机制。市民要求参与到城市建设发展决策过程的热情十分高涨，这种现象不仅仅出现在柏林，德国，欧洲，而是一种全世界发展的潮流趋势，公民越来越积极地参与到与自身休戚相关的重大城市发展决策当中来。如何有效地引导公众参与，这是每个政府所面临的问题。联建住宅项目则为公众参与提供了一个规模与程度都适宜的试炼场地，各种各样的问题都会在项目进行的过程中暴露无疑，于是经验得到积累。通过在众多联建项目中的实践，项目成员的“参与资历”得到增长，在面对规模与程度更大的城市发展建设问题时，能够更加有效有序地参与，最大限度地发挥公众参与所带来的优势与益处。

第三，联建住宅项目整合了零散的城市空间资源，填补修复了破碎的城市街块。Fig.4.3 显示了柏林范围内空地的分布，它们的存在被许多专家认为是内城新一轮发展的催化剂。联建住宅项目多为规模较小的住宅项目，与散落在城市街块之中的小型空地尺度相符，因而它们成为了填补城市街块空缺的重要方式。在本书搜集的 40 个案例中，有接近半数的项目是在嵌入式小型地块上进行的，这些项目小心翼翼地处理与相邻建筑的关系，修复了破裂的城市界面。

4.1.3 柏林联建住宅在形式与空间方面的意义

第一，联建住宅项目发展拓宽了居住空间的形式与功

on the elderly.Besides many Co-housing projects also create safe and comfortable living area for the vulnerable groups including women,children and the sick people; additionally some Co-housing projects will share parts of functions of social welfare institutions by offering relevant organization service space with relatively low rent (Case 17 Suedwestsonne, Case 19 Lebensort Vielfalt and etc.).

Thirdly, Co-housing projects have integrated idle capitals, experimental academic research achievements and other social resources and made full use of them to create platform for urban practice. By gathering the idle money from the group members, Co-housing projects lower the initial capital for the individual to build a house thus allowing more middle and low income groups to participate in the practice of self-made housing projects; meanwhile Co-housing projects have made experiments on housing in small scale possible. Some projects that will never get the chance to practice in normal real estate market can find opportunities in the way of Co-housing. It helps promote the development of housing projects on structure- and material selecting to some extent.

4.1.2 Significance in Urban Design

Firstly, Co-housing projects have supplemented and enriched the forms and contents of urban space, and laid the spatial foundation for the promotion of local sense of community. Co-housing projects have broken up the boundary between public and private space to some degree, and inserted another level of shared space as the transition between the urban level and the small community level; the forms and contents of shared space in Co-housing vary including rest, recreation, education,culture, community service, urban practice and other functions, which radiate

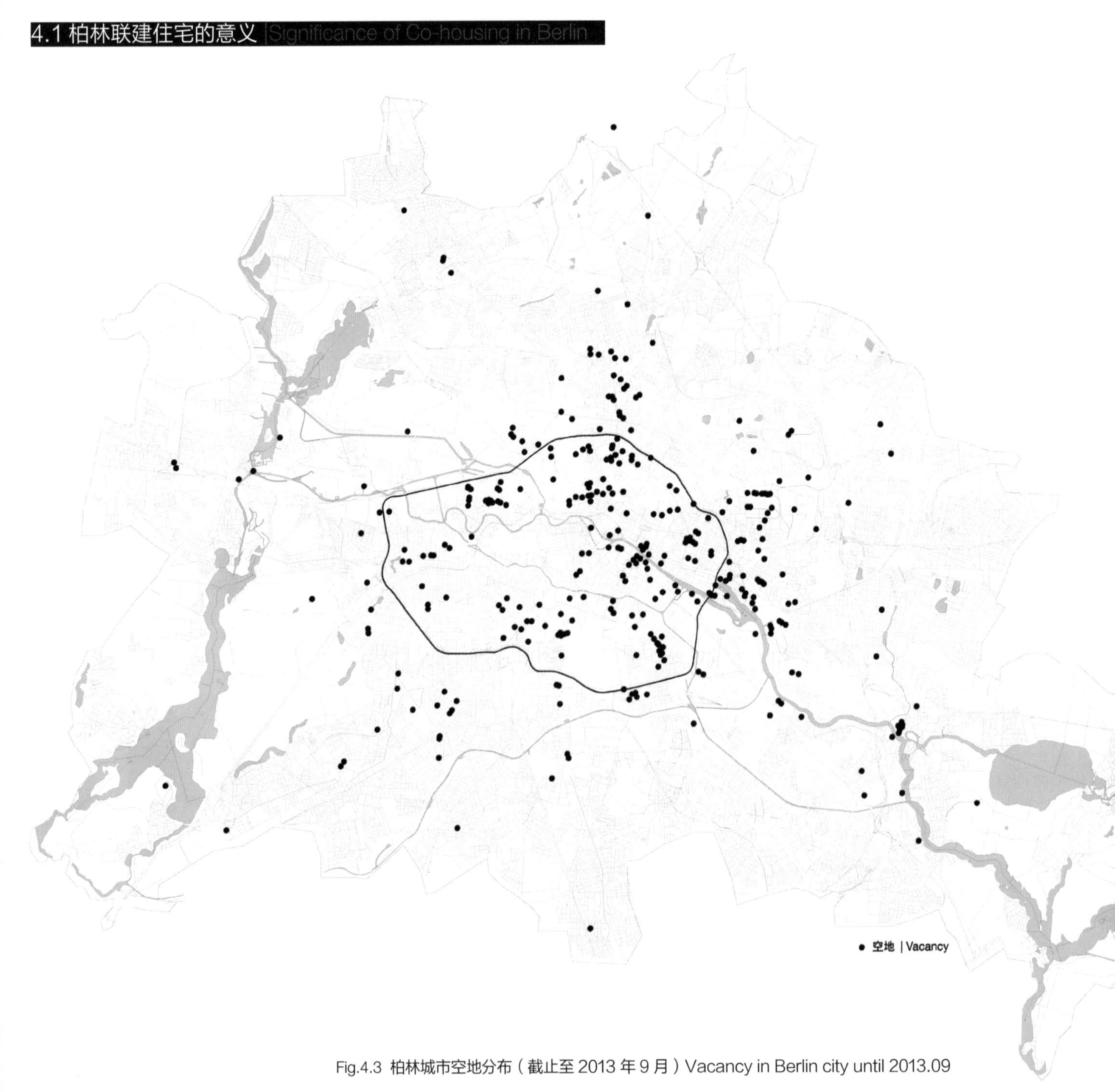

Fig.4.3 柏林城市空地分布（截止至 2013 年 9 月）Vacancy in Berlin city until 2013.09

能，补充发展了在传统房地产开发项目中难以涉及到的户型与空间（特殊地块上的特殊户型）。由于联建住宅项目是由居住者高度订制的，与传统房地产开发项目的普适户型不同，它们往往有更多个性化的设计，甚至某些在他人看来不可能被用于居住的极限空间也会出现在种类繁多的联建住宅项目中；另一方面联建住宅项目也在积极探索住宅可适应性方面的可能性，通过研究灵活性极大的标准户型来满足多变的住户需求。

第二，联建住宅项目探索并研究了共享空间在住宅建筑中可能的形式、功能以及构成关系。作为具有公共属性的共享空间，按照传统的居住理念，是不应该出现在具有私密属性的居住空间之中的。但随着现代居住观念的改变，居住氛围的逐渐恶劣，以及人与人之间的交流越来越局限，越来越多的人渴望与人分享，并且建立互信互助的社区生活关系。由是，基于社区生活的联建住宅项目应运而生，而共享空间作为社区生活的物质基础，成为了重要的存在。这种存在没有被局限在户外，而是延伸到了室内的居住生活空间。联建住宅项目通过实践探索了室内外共享空间的形式与功能，在某种程度上创造了一种新的住宅类型。

第三，联建住宅项目将使用者直接引入设计过程，参与建筑形式与空间的设计与规划，创造出有别源于建筑师个人经验的住房产品。在联建住宅项目中，建筑师根据未来使用者的需求以及自身的经验制定切实可行的“游戏规则”（包括单元模块，组合规则等）但不决定最终的结果，使用者成为玩“游戏”的人，得到整个建筑的大致形态最后在建筑师的润色下形成最终的方案。

4.1.4 柏林联建住宅在可持续与生态方面的意义

第一，联建住宅项目设立了节能环保的新标准，通过项目实践推动了相关法律法规的发展与完善。很多联建住宅项目在相关的节能环保法律条文出来之前，就开始了这方面的探索，将环保节能理念整合进建筑设计之中。

第二，联建住宅项目为其他住宅项目在经济、文化、资源、能源方面的可持续树立了典范，通过广泛地建立合作

the surroundings and become important tools of building up positive and stable social network within the community.

Secondly,Co-housing projects have introduced the future residents and the neighborhood into the decision-making process of urban design. It helps accumulate experience on civic participation and perfect the participating mechanism. The desires of citizens to take part in the decision-making process of important urban development that will have great influences on themselves becomes stronger and stronger. It is not a unique situation in Berlin or Germany but a global trend. That how to lead the public to participate effectively is of importance for every government in different levels. Co-housing projects just provide such training ground for participation of proper scale and degree. Various issues (equity and equality issue, efficiency issue, group member maximum issue, participating frequency issue and etc.) expose during the process, with experience accumulated and lessons learned. Via the practice in the Co-housing projects, group members have gained the knowledge of “participation”. When facing bigger and larger urban development issues, they can participate in order and effectively and benefit from public participation to the most.

Thirdly, Co-housing projects have integrated scattered usable urban space resources and helped repair the broken urban block. Fig.4.3 shows the vacant space in Berlin. In the views of many experts these space resources are the catalysts for the next round of inner city development. Most of Co-housing projects are middle and small-scaled which match well with the sizes of these vacant space, so Co-housing projects have become the important way of “filling blank”. In the selected 40 cases, nearly half of them are located in the embedded vacant space carefully dealing with the relationship with the adjacent buildings to recover the urban interface.

关系网络，更好更有效地整合利用了既有资源，可持续的理念贯穿项目的始终。

第三，联建住宅项目在建筑构成、使用人群构成、居住生活方式、项目理念等方面探索研究了住宅建筑可持续利用的可能性， 拓展了可持续建筑的内涵。由于联建住宅项目的初衷之一就是将住房永久地隔离在市场之外，如何能够使得项目能够永久地持续下去，这涉及到使用人群的更替问题，因而对于联建住宅的可适应性提出了较高的要求。

4.2 柏林联建住宅的局限

柏林联建住宅在具备上述意义的同时，仍然存在着自身的局限性：

第一，效率问题。公众参与是联建住宅项目的一大优势，而反过来这一优势又成为了项目持续高效进展的最大障碍。每个成员都要平等公平地发表意见和建议，对于一个议题要反复地讨论协商，大量的时间花费在了前期的讨论决策上，如 Fig.4.4 所示，前期的筹备阶段（包括图中的 1–3 阶段）往往要经历 2 年的时间，而随着项目规模的增大，这一时间段也被拉长。联建住宅项目的规模是否存在上限、上限为何，这些都需要通过众多的实践才能获知。从目前的情况来说，联建住宅项目更加适合比较小型的团体进行实践，并最好控制在百人以下，四五十人为宜。

第二，地块问题。适宜的中小地块资源的使用已渐趋饱和，如何有效利用大型地块以及联建项目上限规模的确定，也是十分实际的问题；为了避免联建团体内部成员数量的过分庞大，在开发大型地块时，往往是划分为若干个小型联建团体共同开发，或是之前形成的小型团体联合起来共同开发大型地块。但是对于过大的地块这种方式也并不一定能够奏效，仍然需要通过实践检验出该种方式的上限。

第三，未来走向问题。联建住宅项目由第一代转入第二代项目的可持续性问题，项目如何保持初衷稳定地维持下去，这将是所有项目面临的较长远的问题。

4.1.3 Significance in Form and Space

Firstly, Co-housing projects have enriched the form and function of living space and supplemented the types that are hardly involved in the normal real estate housing projects. Because Co-housing projects are highly customized, many individualized “extreme” living space can be applied in the housing projects. They are exploring the possible “non-standardized” housing forms and space; what’s more, Co-housing projects are also exploring the possibilities of housing on adaptability.

Secondly, Co-housing projects have studied the form, function and spatial constitution of shared space in housing. As public space, shared space is not supposed to exist in the private space like living space, but with the development of modern living concept and the lack of social communication, more and more people hope to share with each other and to build up a living model based on community support. Co-housing comes into being and the shared space becomes a vital existence as the basis of community life. Shared space exists both indoor and outdoor. Practicing Co-housing projects are exploring the possible form and function of shared space, thus creating a new housing typology to some degree.

Thirdly, Co-housing projects have introduced users into the process of form and space design and created housing products based on the users’ demands instead of experience from the architects. Usually in Co-housing projects, the architects formed the “game rules” based on their experience and the demands of the future users but didn’t decide the final results. The users became the “game players” and got the almost final results. Then after the “embellishment” of architects the whole group would get the final one.

4.1.4 Significance in Sustainability and Ecology

Firstly, Co-housing projects have set new standards on energy-saving and environmental protection, thus promoting the development and improvement of relevant laws and regulations to some extent. Many Co-housing projects started the related exploration and integrated the concept into the architecture design long before the establishment of the relevant laws and regulations on energy-saving and environmental protection.

Secondly, Co-housing projects set examples for other housing projects on the sustainability of economy,culture, resource and energy. Through extensive cooperation social network, Co-housing projects integrate and utilize the existing resources better and more effectively.

Thirdly, Co-housing projects are exploring the possibilities of housing sustainability on spatial constitution, user composition, living style and project concept and extend the content of sustainable buildings. Since one of the original intention of many Co-housing projects is to exclude housing permanently from the speculating market. That how to sustain the projects permanently refers to the issue of user group's replacement, which asks for high requirements on the housing adaptability.

4.2 Limitation of Co-housing in Berlin

Although Co-housing in Berlin shares the significance above, they still have their limitations:

Firstly, Efficiency Issue. Participation is one of the advantages of Co-housing projects but it can also be the big problem hindering the continuous and effective progress of the project the other way around. Each member need to express his opinions equally and the same topic needs discussion again and again. Much time has been spent on the preliminary decision-making process. As Fig.4.4 shows the preliminary preparation period (including 1-3 phases in Fig.4.4) usually lasts 2 years and the scale increases this period will be prolonged accordingly.Is there any maximum limit for the scale of the Co-housing group? If so, what is it? All these questions need to be answered through the practice. For the current situation, Co-housing projects are more suitable for small- and middle-scaled group within 40-50 people.

Secondly, Site Issue. Proper site resources of middle and small sizes are nearly exhausted. That how to utilize the large open site and how to define the maximal scale of the project are realistic issues; in order to avoid excessive number of group members, usually the large site is divided into smaller sites for several small Co-housing groups to develop. If the site is excessive large, usually the way may not be workable so the maximal limit of site also need to be figured out through practice.

Thirdly, Succession Issue. That how to sustain the project to the second generation, third generation and even longer is the issue that all the Co-housing projects will face.

阶段 Phases	活动 Activities		持续时间 Duration
1 动员阶段 Orientation Phases	· 利益相关团体 · 寻找成员 · 寻找专家 · 寻找场地 · 概念草图 · 预算 · 组织及法律形式	Interest Group Search for Further Members Search for Experts Search for Buildings or Site Conceptual Design Outline cost estimates Organisational & Legal Form	6-12 个月 6-12 Months
2 规划阶段 Planning Phases	· 规划法规要求 · 选择专业建筑师与规划师 · 选择项目经理 · 确定场地 · 规划建造项目 · 优化设计 · 造价计算 · 确保资金	Planning Regulations Requirements Selection of Architect/ Specialist Consultants Selection of Project Manager Real Estate Purchase Option Construction Project Planning Optimise Design Cost Calculations Secure the Financing	3-6 个月 3-6 Months
3 购买和建造筹备阶段 Purchase & Construction Preparation Phases	· 购买场地 · 建造活动目录 · 招标 · 工程保险 · 签订合同	Purchase the Plot Bill of quantities of Construction Work Tendering for Construction Contracts Construction Insurance Award of Contract	4-6 个月 4-6 Months
4 建造阶段 Construction Phases	· 监管 · 资金控制 · 结算 · 验收	Site Management/Supervision Financial Control Accounting of Modernisation/ Own Contribution/ Construction Final Acceptance	12-15 个月 12-15 Months
5 居住阶段 Residential Phases	· 监督管理 · 建筑管理 · 组内活动	Care & Managementn Building Management Group Process	

Fig.4.4柏林联建住宅的一般组织过程 General Organization process of Co-housing projects in Berlin

龚喆
Gong Zhe

Fig.1.1 联建住宅的特征 |Co-housing's Characteristics
Fig.1.2 联建住宅的利益相关者 | Actors of Co-housing
Fig.1.3 德国联建住宅发展时间轴 | Timeline of Co-housing's development in Germany
Fig.1.7 柏林联建住宅项目分布图（截止至 2012 年十月）| Co-housing in Berlin until 2012.10

Fig.2.1 研究案例分布图 | Studied Cases'Distribution
Fig.2.2 柏林联建住宅类型 | Co-housing's typologies in Berlin
Fig.2.3 40 个研究案例按照委托方、场地类型、空间组织模式分类
| 40 Studied Cases are classified according to Clients, Site types & Space organisation Models

Fig.3.1 参与性与共享性 |Participation & Sharability
Fig.3.2 项目区位及实景（区位部分）| Project location & scene (Location Part)
Fig.3.3 Spree Berlin 共享空间体系示意图 Shared space system of Spree Berlin
Fig.3.4 项目区位及实景（区位部分）| Project location & scene (Location Part)
Fig.3.6 共享空间体系示意图 |Shared communal space system
Fig.3.7 项目区位及实景（区位部分）| Project location & scene (Location Part)
Fig.3.9 共享空间分析 | Shared space analysis
Fig.3.10 项目区位及实景 |Project location & scene
Fig.3.12 项目区位及实景（区位部分）|Project location & scene (Location Part)
Fig.3.13 户外活动发生频率与空间质量的关系 | relation between possibilities of activities and space qualities
Fig.3.14 参与性与共享性 | participation & sharability
Fig.3.15 不同性质空间之间的空间关系 |relationship between different kinds of space
Fig.3.16 共享空间嵌入手法：作为并列空间 |ways of inserting shared space：as juxtaposed space
Fig.3.17 共享空间嵌入手法：作为从属空间 |ways of inserting shared space：as subordinate space
Fig.3.18 订制性与个性 | Customization & Identity
Fig.3.19 项目区位及实景（区位部分）|Project location & scene (Location Part)
Fig.3.20 项目区位及实景（区位部分）|Project location & scene (Location Part)
Fig.3.22 项目区位及实景（区位部分）|Project location & scene (Location Part)
Fig.3.23 项目日常运营收支示意图 |Project daily incomings and outcomings demonstration
Fig.3.24 项目区位及实景（区位部分）|Project location & scene (Location Part)
Fig.3.27 项目区位及实景（区位部分）|Project location & scene (Location Part)
Fig.3.29 项目区位及实景（区位部分）|Project location & scene (Location Part)
Fig.3.33 项目区位及实景（区位部分）|Project location & scene (Location Part)
Fig.3.37 联建住宅与传统住宅在订制性与个性上的区别 |Difference between Co-housing and traditional housing on Customization & Identity
Fig.3.38 个性化生活方式的探索 | Exploration on Individualized living style
Fig.3.39 个性化户型设计策略 |Individualized housing unit design strategy
Fig.3.40 个性化户型组织形式 |Individualized housing unit organization forms
Fig.3.41 生态性与可持续性 |Ecology & Sustainability
Fig.3.42 项目区位及实景（区位部分）|Project location & scene (Location Part)
Fig.3.44 项目区位及实景（区位部分）|Project location & scene (Location Part)
Fig.3.47 项目区位及实景（区位部分）|Project location & scene (Location Part)
Fig.3.48 项目发展及其建构 | Project's development and its tectonic
Fig.3.49 项目区位及实景（区位部分）|Project location & scene (Location Part)
Fig.3.50 项目区位及实景（区位部分）|Project location & scene (Location Part)
Fig.3.51 项目区位及实景（区位部分）|Project location & scene (Location Part)
Fig.3.53 联建住宅与传统住宅在生态性与可持续性上的区别 | Difference between Co-housing and traditional housing on Ecology & Sustainability
Fig.3.54 项目可持续性的实现策略 |Strategy to realize Project Sustainability
Fig.3.55 自发性与探索性 | Spontaneity & Exploration
Fig.3.56 项目区位及实景（区位部分）|Project location & scene (Location Part)
Fig.3.59 项目区位及实景（区位部分）|Project location & scene (Location Part)
Fig.3.60 项目区位及实景（区位部分）|Project location & scene (Location Part)
Fig.3.61 项目区位及实景（区位部分）|Project location & scene (Location Part)
Fig.3.62 项目区位及实景（区位部分）|Project location & scene (Location Part)
Fig.3.63 项目区位及实景（区位部分）|Project location & scene (Location Part)
Fig.3.64 联建住宅与传统住宅在自发性与探索性上的区别 |Difference between Co-housing and traditional housing on Spontaneity & Exploration
Fig.3.65 项目成本控制的实现策略 |Strategy to realize Project cost control

Fig.4.3 柏林城市空地分布（截止至 2013 年 9 月）|Vacancy in Berlin city until 2013.09

ktualisierte und
rweiterte Ausabe:
rojektuebersicht
nternationale
auausstellung
erlin 1987

Fig.1.4 1984-1987 柏林国际建筑展旧建筑改造项目分布图 |Internationale Bauausstellung Berlin 1984-1987 urban renewal project map

图片来源 |Image Source

Möckernkiez e.G.	Fig.1.5 慕肯社区规划讨论会 |Möckernkiez e.G. planning workshop
M. Fedrowitz	Fig.1.6 2010 年德国联建住宅项目分布图 |Co-housing project map in Germany until 2010
Netzwerkagentur Generationen Wohnen	Fig.1.8 柏林联建住宅三种所有制模式 |Co-housing's ownership Fig.1.9 柏林联建住宅项目的实现形式 |Co-housin's approach Fig.4.4 柏林联建住宅的一般组织过程 |General organisation process of Co-housing projects in Berlin
Tristan Thönnissen	Fig.3.2 项目区位及实景（实景部分）|Project location & scene（Scene Part）
Marcus Thiele Andrea Kroth	Fig.3.4 项目区位及实景（实景部分）|Project location & scene（Scene Part） Fig.3.27 项目区位及实景（实景部分）|Project location & scene（Scene Part）
Zanderroth Architekten	Fig.3.5 实景 |Scene Fig.3.28 灵活的平面 |Flexible plan
Fatkoehl Architekten	Fig.3.7 项目区位及实景（实景部分）|Project location & scene（Scene Part） Fig.3.8 项目街块示意图 |Block demonstration
R50_ifau_HVB_Wohn-reporte	Fig.3.11 共享空间体系示意图 |Shared communal space system Fig.3.30 居住空间评估参考体系 | Methodological evaluation of residential spatial reference systems Fig.3.31 项目模数系统 |Project module system Fig.3.32 各层平面及居民自由选择模块所形成的立面效果 | Plan for Each Floor and Individualized Facade by module system
Carpaneto Schoeningh Architekten	Fig.3.12 项目区位及实景（实景部分）|Project location & scene（Scene Part）
Christian Muhrbeck	Fig.3.19 项目区位及实景（实景部分）|Project location & scene（Scene Part）
Baufrösche Architects and Urban Planners GmbH	Fig.3.20 项目区位及实景（实景部分）|Project location & scene（Scene Part） Fig.3.21 项目总平面图 | Master Plan
M29 Hausprojekt	Fig.3.22 项目区位及实景（实景部分）|Project location & scene（Scene Part）
Jan Bitter	Fig.3.24 项目区位及实景（实景部分）|Project location & scene（Scene Part）
BAR Architekten	Fig.3.25 空间组织关系示意图 |Space organization demonstration
Kleinein, D.	Fig.3.26 项目所有制模式示意图 |Project ownership model demonstration
Andrew Alberts	Fig.3.29 项目区位及实景（实景部分）|Project location & scene（Scene Part） Fig.3.56 项目区位及实景（实景部分）|Project location & scene（Scene Part）
Klemens Ortmeyer Ludger Paffrath	Fig.3.33 项目区位及实景（实景部分）|Project location & scene（Scene Part）
Deadline Architekten	Fig.3.34 项目改造前实景 |Scene before renewal Fig.3.35 项目改造后实景 |Scene after renewal Fig.3.36 项目平面及剖面 Project plan & section
Bernd Borchardt	Fig.3.42 项目区位及实景（实景部分）|Project location & scene（Scene Part）
Kaden-Klingbeil Ar-chitekten	Fig.3.43 项目剖面及平面图 |Project section & plan Fig.3.60 项目区位及实景（实景部分）|Project location & scene（Scene Part）

Source	Figure
Stefan Mueller David von Becker	Fig.3.44 项目区位及实景（实景部分）|Project location & scene（Scene Part）
Fertighauscity 5+	Fig.3.45 木构研究模型 |Woodstructure research model
Institut fuer Urbanen Holzbau	Fig.3.46 项目各层平面图 |Project plan for each floor
Frank Huelsboemer	Fig.3.47 项目区位及实景（实景部分）|Project location & scene（Scene Part） Fig.3.59 项目区位及实景（实景部分）|Project location & scene（Scene Part）
Deimel Oelschlaeger Architekten	Fig.3.49 项目区位及实景（实景部分）|Project location & scene（Scene Part）
Wilde Gaertnerei	Fig.3.50 项目区位及实景（实景部分）|Project location & scene（Scene Part）
www.louisp.de.	Fig.3.51 项目区位及实景（实景部分）|Project location & scene（Scene Part） Fig.3.52 能源利用概念 |Energy concept
Heide & von Beckerath Architekten	Fig.3.57 基于标准模块的个性设计 |Individualized design based on standardized module Fig.3.58 项目底层平面图 |Project groundfloor
Archid Architektur	Fig.3.61 项目区位及实景（实景部分）|Project location & scene（Scene Part）
Mueggelhof Project	Fig.3.62 项目区位及实景（实景部分）|Project location & scene（Scene Part）
Schwulenberatung Berlin	Fig.3.63 项目区位及实景（实景部分）|Project location & scene（Scene Part）
IDN ImmoDaten IBB_ Wohnungs marktbericht_ 2013	Fig.4.1 2005-2013 年柏林出租房基础租金变化 |basic rent changes in Berlin from 2005-2013 Fig.4.2 2005-2013 年柏林私有住房售价变化 |Housing purchasing price changes in Berlin from 2005-2013

著作 | Books：

[01] Ahrensbach, T., Beunderman, J., Fung, A., Johar, I. and Steiner, J. (2011) Compendium for the Civic Economy, London: Calverts Co-operative.
[02] Buhtz, M., Gerth, H., Lindner, M. and Marsch, S. (2009) Familienwohnen in der Stadt:
Beispiele fuer kinder-und familienfreundliches Bauen und Wohnen in der Berliner Innenstadt [online], available:
http://www.stadtentwicklung.berlin.de/wohnen/familienfreundlich/download/broschuere_famfreundwohnberlin2010.pdf [accessed 10 Jan 2014]
[03] Cremer, C., Nikolaus, M., Pfander, H. and Praum, C. (2012) Wohnen in Gemeinschaft: Von der Idee zum gemeinsamen Haus [online],
available: http://www.netzwerk-generationen.de/fileadmin/user_upload/PDF/Downloads_brosch%C3%BCren-dokumentationen/B5_Broschuere_Stattbau_neu_fertig-web.pdf
[accessed 12 Sep 2013]
[04] Hamdi, N., Goethert, R. (1997) Action Planning for Cities: A Guide to Community Practice, England: John Wiley & Sons Ltd.
[05] Heyden, M., comp. (2007) Berlin: Wohnen in eigener Regie, Berlin: Agit-Druck.
[06] id22, (2012) Cohousing Cultures : Handbook for self-organized, community oriented and sustainability housing, Berlin: Jovis Varlag GMBH.
[07] LaFond, M. (2011) Cohousing cultures are the future of social housing: Planning, Designing and managing projects in Berlin and Europe, Proceeding International conference on social housing seminar in Taiwan; The Right to Adequate Housing and Social Inclusion, 176
[08] LaFond, M., Honeck, T., Suckow, C. (2012) Self-organized, community oriented, sustainable; Cohousing Cultures; Handbook for self-organized, community oriented and sustainability housing, Jovis Varlag Gmbh Berlin
[09] Mindak, J., comp. (1987) Step by Step: Careful urban renewal in Kreuzberg, Berlin: STERN
[10] McCamant, M., K., Dorrit, C., R. (1988) Cohousing: A contemporary approach to Housing Ourselves, California USA: Habitat Press.
[11] Overmeyer, K. (2007) Urban pioneers: Berlin Stadtentwicklung durch Zwischennutzung, Berlin: Jovis-Verl.
[12] Ring, K., Eidner, F., Merker, J. and Sawal A. (2007) Auf Einander Bauen, Berlin: Druckerei Lokay.
[13] Ring, K. (2013) Selfmade City Berlin: Stadtgestaltung und Wohnprojekte in Eigeninititie, Berlin: JOVIS.
[14] Scotthanson, C. and Scotthanson, K. (2005) The Cohousing Handbook: Building a Place for Community, Canada: New Society Publisher.
[15] Senstadt (2012) IBB Housing Market Report 2012, Senatsverwaltung für Stadtentwicklung und Umwelt Berlin
[16] Walz, S., Kast, A., Schulze, G., Born, L., Krueger, K. and Niggemeier, K. (2011) Handbuch Zur Partizipation, Berlin: Trigger Medien GMBH

[17] (日) 芦原义信 . 外部空间设计 [M]. 尹培桐译 . 北京 : 中国建筑工业出版社 ,1985.
[18] 彭一刚 . 建筑空间组合论 [M]. 北京： 中国建筑工业出版社 ,1998.
[19] (德) 阿 · 考夫卡 ,(美) 温迪 · 科恩编. 柏林建筑 [M]. 张建华 , 杨丽杰译 . 辽宁： 辽宁科学技术出版社 ,2000.
[20] (美) 凯文 · 林奇 . 城市意象 [M]. 方益萍，何晓军译 . 北京： 华夏出版社 ,2001.
[21] (丹麦) 扬 · 盖尔著 . 交往与空间 [M]. 何人可译 . 北京 : 中国建筑工业出版社 ,2002.
[22] (德)H.M. 纳尔特 . 德国新建筑 [M]. 杨宇宁等译 . 大连 : 大连理工大学出版社 ,2002.
[23] 汪丽君，舒平 . 类型学建筑 [M]. 天津 : 天津大学出版社 ,2003.
[24] 李振宇 . 城市 · 住宅 · 城市： 柏林与上海住宅建筑发展比较 [M]. 南京 : 东南大学出版社 ,2004.
[25] 沈祉杏 . 穿墙故事： 再造柏林城市 [M]. 北京 : 清华大学出版社 ,2005.
[26] [美] 彼得 · 艾森曼 . 彼得 · 艾森曼图解日志 [M]. 陈欣欣 , 何捷译 . 北京 : 中国建筑工业出版社 ,2006.
[27] [意] 阿尔多 · 罗西 . 城市建筑学 [M]. 黄士钧译 . 刘先觉校 . 北京 : 中国建筑工业出版社 ,2006.
[28] 李振宇 , 邓丰 , 刘智伟 . 柏林住宅： 从 IBA 到新世纪 [M]. 北京 : 中国电力出版社 ,2007.
[29] [美] 道格拉斯 · 凯尔博 . 共享空间——关于邻里与社区设计 [M]. 吕斌 , 覃宁宁 , 黄翊译 . 北京 : 中国建筑工业出版社 ,2007.
[30] 周静敏 . 世界集合住宅： 都市型住宅设计 [M]. 北京 : 中国建筑工业出版社 ,2011.

期刊 | Journal Articles：

[31] Arnstein, S. R. (1969) ‘A Ladder of Citizen Participation’, JAIP, 35(4), 216-224.
[32] Ache, P., and Fedrowitz, M. (2011) ‘The development of Co-housing Initiatives in Germany’, Built Environment, 37(3), 395-412.
[33] Ballhausen, N. (2006) ‘Anklamer Strasse 52: Ein Wohnhaus fuer eine Baugruppe in Berlin’,
Bauwelt, 03/03, 16-19.
[34] Ballhausen, N. (2008) ‘Bernauer Straße 5d - Haus FL’, Bauwelt, 24/10, 36-37. [17] Krokfors, K. (2011)‘Co-housing in the Making’, Built Environment, 38(2), 309-314.
[35] Ballhausen, N. (2008) ‘Strelitzer Straße 53: vom Schussfeld zum Bauland’, Bauwelt, 24/10, 28-33.
[36] Ballhausen, N. (2012) ‘Den Holzbau radikalisieren’, Bauwelt, 25/05, 36-41.
[37] Brinkmann, U. (2012) ‘Geschosswohnungsbau: Gegenüber der Gleisbrache’, Bauwelt, 09/03,14-19.
[38] Fedrowitz, M. and Gailing, L. (2003) ‘Zusammen wohnen: GemeinschaftlicheWohnprojekte als Strategie sozialer und ökologischer Stadtentwicklung’, Dortmunder Beiträge zur Raumplanung, No. 112.
[39] Fedrowitz, M. (2011a) ‘Gemeinschaftliches Wohnen in Deutschland’ in Nationalatlas 9. Leipzig: Leibniz-Institut für Länderkunde (IfL) [online]. Available:
http://aktuell.nationalatlas.de/wohnprojekte-9_09-2011-0-html. [accessed 11 Sep 2013]
[40] Fedrowitz, M. (2011b) ‘Vernetzung von Wohn-projekten mit ihrem Quartier’ in wohnbunde.V. (Ed.) Perspektiven für Wohnprojekte. wohnbund-informationen, No.1, 42–44.
[41] Kleinein, D. (2006) ‘Professioneller Selbstbau: Neue Ökonomien in der Projektentwicklung’, Bauwelt, 03/03, 30-33.
[42] Kleinein, D. (2008) ‘Zwillingshäuser Sc10, Ru43: Der Städtebau des Machbaren’, Bauwelt, 24/10, 26-27.
[43] Kleinein, D. (2008) ‘Esmarchstraße 3: Kritische Verkapselung’, Bauwelt, 18/04, 18-23.

[44] Kleinein, D. (2009) 'Brunnenstraße 9:Der inszenierte Rohbau', Bauwelt, 11/12, 12-21.
[45] Kleinein, D. (2010) 'Slow Architecture', Bauwelt, 05/11, 14-21.
[46] Kleinein, D. (2011) 'BIGYard: Die größte Baugruppe Berlins', Bauwelt, 29/07, 12-21.
[47] Ortmeyer, K. (2003) 'Hessische Straße 5 - Slender', Bauwelt, 10/01, 20.

[48] 李振宇，虞艳萍 . 欧洲集合住宅的个性化设计 [J]. 中外建筑 , 2004,3:3-8.
[49] 李振宇，刘智伟 .IBA 新建内城住宅的设计启示 [J]. 建筑师，2004,2:29-33.
[50] 李振宇 . 欧洲住宅建筑发展的八点趋势及其启示 [J]. 建筑学报 , 2005,4:78-81.
[51] 杨田 . 居住区户外交往空间与邻里关系的思考 [J]. 南京艺术学院学报：美术与设计版 , 2010,2:147-149.
[52] 薄力之 , 邓琳爽 . 边缘社区的公众参与 - 以柏林邻里管理项目为例 [J]. 城市研究，2013,2:88-93.

学位论文 | Thesis and Dissertations：

[53] Sudiyono, G. (2013) Learning from the management and development of current self-organized and community-oriented Cohousing projects in Berlin: making recommendations for the IBA Berlin 2020, unpublished thesis (M.A.), TU Berlin
[54] Rauscher, T. (2013) Work, Community and Sustainability-Redefing Work through Cohouisng, unpublished thesis (M.A.), Lund University

[55] 张睿 . 合作建房国外经验借鉴和我国相关制度的建构 [D]. 天津：天津大学建筑学院 ,2007.
[56] 张睿 . 国外合作居住社区研究 [D]. 天津 : 天津大学建筑学院 ,2011.
[57] 李晓蕾 . 合作居住的设计理念及应用研究 [D]. 天津 : 天津大学建筑学院 ,2007.
[58] 虞艳萍 . 当代欧洲集合住宅个性化设计的特征与手法研究 [D]. 上海 : 同济大学 ,2005.
[59] 王芳 . "谨慎的城市更新" 柏林国际建筑展 (IBA，1984-87) 旧区住宅更新研究 [D]. 上海：同济大学，2004.
[60] 邓丰 . 柏林国际建筑展 (IBA，1984-1987) 新建住宅外部空间研究 [D]. 上海：同济大学 ,2004.
[61] 沈芋芋 . 柏林：建筑与城市设计的理念实验场—政治因素与城市建设的互动之初探 [D]. 南京：东南大学 ,2005.

序号 No.	项目名称 Name	地址 Address	类型 Typology	设计方 Designer	成立时间 Time	图片资料 Figure
01	Spreefeld Berlin	Köpenicker Strasse 48/49	居民委托型，开敞式场地，集群式 Resident Open Field Cluster	ARGE SILVIA CARPARNETO, FAT KÖHL, BAR ARCHITEKTEN	2010	
02	Zwillingshäuser	Schönholzer Str.10A	居民委托型，嵌入式场地，标准单元重复式 Resident Embedded Field Standardized Unit	ZANDERROTH ARCHITEKTEN	2005	
03	Baugemeinschaft Simplon	Simplonstrasse 54	建筑师自我委托型，修复改扩建式，标准单元重复式 Architect Renovated Field Standardized Unit	FAT KOEHL ARCHITEKTEN	2008	
04	R50	Ritterstrasse50	建筑师自我委托型，开敞式场地，标准单元重复式 Architect Open Field Standardized Unit	IFAU+JESKO FEZER HEIDE & VON BECKERATH ARCH	2010	
05	Wohnetagen Steinstrasse	Steinstrasse 26-28	建筑师自我委托型，修复改扩建式， 非标准单元组合式 Architect Renovated Field Non-Standardized Unit	CARPANETO SCHÖNINGH ARCHITEKTEN	1998	
06	Allein Wohnen in Gemeinschaft	Falkstrasse 25	政府委托型，修复改扩建式，标准单元重复式 Government Renovated Field Standardized Unit	N/A	2005	
07	Möckernkiez EG	Möckernstrasse Yorckstrasse 24	居民委托型，开敞式场地，集群式 Resident Open Field Cluster	DREES+SOMMER, BE BERLIN, R. ROEDIG. SCHOP ARCH.	2007	
08	M29	Malmoeer strasse 29	非营利性组织或团体委托型，开敞式， 标准单元重复式 Non-ProfitEntity Open Field Standardized Unit	CLEMENS KRUG ARCHITEKTEN	2008	
09	Baugruppe Oderberger Strasse 56	Oderberger Strasse 56	建筑师自我委托型，嵌入式场地，非标准单元组合式 Architect Embedded Field Non-Standardized Unit	BAR ARCHITEKTEN	2003	
10	Slender+Bender	Hessische Strasse 5	建筑师自我委托型，嵌入式场地，非标准单元组合式 Architect Embedded Field Non-Standardized Unit	DEADLINE	1999	

序号 No.	项目名称 Name	地址 Address	类型 Typology	设计方 Designer	成立时间 Time	图片资料 Figure
11	E3	Esmarchstrasse 3	建筑师自我委托型，嵌入式场地，标准单元重复式 Architect Open Field Standardized Unit	KADEN-KLINGBEIL ARCHITEK-TEN	2006	
12	3Xgrün	Görschstrasse 48/49	建筑师自我委托型，嵌入式场地，标准单元重复式 Architect Embedded Field Standardized Unit	ARGE ATELIER PK, ROEDIG. SCHOP, INSTITUT FUER UR-BANEN HOLZBAU (IFUH)	2009	
13	Hegemonietempel	Uferstrasse 6	居民委托型，修复改扩建式，标准单元重复式 Resident Renovated Field Standardized Unit	CHRISTOF MAYER, BUERO FUER AR-CHITEKTUR UND STAEDTEBAU	2009	
14	Living in Urban Units	Schönholzer Str.13-14	建筑师自我委托型，嵌入式场地，标准单元重复式 Architect Embedded Field Standardized Unit	PLAANPOOL ARCH. MIT WERKGRUPPE KLEIN-MACHNOW	2006	
15	Louis P.	Pasteurstrasse 27	建筑师自我委托型，嵌入式场地，非标准单元组合式 Architect Embedded Field Non-Standardized Unit	PLAANPOOL ARCH. MIT WERKGRUPPE KLEIN-MACHNOW	2008	
16	Flottwellstrasse 2	Flottwellstrasse 2	建筑师自我委托型，嵌入式场地，非标准单元组合式 Architect Embedded Field Non-Standardized Unit	HEIDE & VON BECKER-ATH ARCHITEKTEN	2007	
17	Südwestsonne	Scharnweber strasse 45	建筑师自我委托型，嵌入式场地，标准单元重复式 Architect Embedded Field Standardized Unit	ARCHID ARCHITEKTUR	2007	
18	Müggelhof Fried-richshain	Müggelstrasse 21	非营利性团体或组织委托型，嵌入式场地，标准单元重复式 Non-Profit Entity Embedded Field Standardized Unit	STEFANIE RUHE	2008	
19	Lebensort Vielfalt	Niebuhrstrasse 59/60	非营利性团体或组织委托型，修复改扩建式，标准单元重复式 Non-Profit Entity Renovated Field Standardized Unit	ROEDIG SCHOP AR-CHITEKTEN	2007	
20	Urbane Living 01	Joachimstrasse 5	建筑师自我委托型，嵌入式场地，非标准单元组合式 Architect Embedded Field Non-Standardized Unit	ABCARIUS+BURNS ARCHITECTURE DESIGN ABCARIUS+ BURNS ARCHITEC-TURE DESIGN	1996	

序号 No.	项目名称 Name	地址 Address	类型 Typology	设计方 Designer	成立时间 Time	图片资料 Figure
21	August 51	Auguststrasse 51	建筑师自我委托型，嵌入式场地，集群式 Architect Embedded Field Cluster	GRUENTUCH ERNST ARCHITEKTEN	2004	
22	Weg Linienstrasse 23	Linienstrasse 23	建筑师自我委托型，开敞式场地，非标准单元组合式 Architect Open Field Non-Standardized Unit	BCO ARCHITEKTEN	2009	
23	Ten In One	Anklamer Strasse 52	建筑师自我委托型，嵌入式场地，非标准单元组合式 Architect Embedded Field Non-Standardized Unit	ROEDIG SCHOP ARCHITEKTEN GBR	2003	
24	Baugemeinschaft Strelitzer Strasse	Strelitzer Strasse 53	建筑师自我委托型，嵌入式场地，非标准单元组合式 Architect Embedded Field Non-Standardized Unit	ANNA VON GWINNER, FLORIAN KOEHL ANDREAS STAHL SDU ARCHITEKTEN	2005	
25	Baugemeinschaft CHO 53	Choriner Strasse 53	建筑师自我委托型，嵌入式场地，标准单元重复式 Architect Embedded Field Standardized Unit	ZOOM ARCHITEKTEN	2005	
26	SC11	Schönholzer Str.11	居民委托型，嵌入式场地，标准单元重复式 Resident Embedded Field Standardized Unit	ZANDERROTH ARCHITEKTEN	2006	
27	Haus FL	Bernauer Strasse 5D	建筑师自我委托型，开敞式场地，集群式 Architect Open Field Cluster	LUDLOFF+ LUDLOFF ARCHITEKTEN BDA	2005	
28	Leuchtturm E.G.	Pappelallee 43	政府委托型，开敞式场地，标准单元重复式 Government Open Field Standardized Unit	MOHR+ WINTERER	2004	
29	Baugemeinschaft AFR 25	Am Friedrichshain 25	建筑师自我委托型，嵌入式场地，标准单元重复式 Architect Embedded Field Standardized Unit	ZOOM ARCHITEKTEN	2006	
30	Big Yard	Zelter Strasse 5	居民委托型，开敞式场地，标准单元重复式 Resident Open Field Standardized Unit	ZANDERROTH ARCHITEKTEN	2006	

序号 No.	项目名称 Name	地址 Address	类型 Typology	设计方 Designer	成立时间 Time	图片资料 Figure
31	Görschstrasse 17	Görschstrasse 17	建筑师自我委托型，开敞式场地，集群式 Architect Open Field Cluster	DMSW ARCHITEKTEN	2010	
32	Meyerbeerstrasse 32	Meyerbeerstrasse 32	建筑师自我委托型，嵌入式场地，标准单元重复式 Architect Embedded Field Standardized Unit	DMSW JULIA DAHL-HAUS	2007	
33	Baugemeinschaft Wohnen am Weissen See GBR MBH	Albertinenstrasse 6-10	建筑师自我委托型，修复改扩建式，集群式 Architect Renovated Field Cluster	ARNOLD GLADISCH ARCHITEKTEN DMSW ARCHITEKTEN STAHLDENNINGER ARCHITEKTEN	2008	
34	Sunny Site Weissensee	Bizetstrasse 65	政府委托型，嵌入式场地，标准单元重复式 Government Embedded Field Standardized Unit	PFEIFER DEEGEN ARCHITEKTEN	2010	
35	Baugruppe K20 GBR	Kreutzigerstrasse 20	居民委托型，嵌入式场地，标准单元重复式 Resident Embedded Field Standardized Unit	ROEDIG SCHOP AR-CHITEKTEN GBR	2005	
36	Werkpalast Lichtenberg	Alfred-Jung-Strasse 8	居民委托型，修复改扩建式，标准单元重复式 Resident Renovated Field Standardized Unit	ANITA ENGELMANN PETER WALTER FALK STELLDINGER	2008	
37	PG Haus 3-15 am Ur-ban (Am Urban GMBH & CO KG)	Grimmstrasse 10, 12-16	建筑师自我委托型，修复改扩建式，集群式 Architect Renovated Field Cluster	GRAETZ ARCHITEK-TEN MBH	N/A	
38	Genowo E.G. Bauherr, Lausitzer Strasse 38	Lausitzer Strasse 38	居民委托型，修复改扩建式，标准单元重复式 Resident Renovated Field Standardized Unit	STADTOASEN AR-CHITEKTEN BUERO	2008	
39	Elffreunde	Vicki-Baum-Strasse 28-40	建筑师自我委托型，开敞式场地，标准单元重复式 Architect Open Field Standardized Unit	AFF ARCHITEKTEN GMBH	2010	
40	Studentendorf Schlachtensee	Wasgenstrasse 75	居民委托型，修复改扩建式，集群式 Resident Renovated Field Cluster	FEHLING, GOGEL & PFANKUCK,AUTZEN & REIMERS, BRENNE ARCHITEKTEN	2003	

本书是作者在双语硕士论文的基础上修改完善而成的，在答辩后的一年时间里，根据中德双方导师的意见和建议，进行了必要的补充和完善。本书亦是同济大学与柏林工业大学双学位联合培养项目的首个出版物，希望它能抛砖引玉，让更多的联培双学位研究成果与人分享。

自 2006 年入读同济大学建筑系，受到学院包豪斯教学模式以及国际化视野的影响，我对德国建筑产生了浓厚的兴趣。在 2009 年学院组织的与瑞士伯尔尼工业大学暑期交流活动中，我有幸结识了后来的研究生导师李振宇教授，正是他引领我跨入了中外住宅建筑研究与探索的大门，李教授始终的支持与鼓励是完成此书的重要动力。

2012 年，我入选了同济大学与柏林工业大学城市设计硕士学位联合培养项目，该项目是学院最早的联合培养项目之一，同时也是李振宇老师倾注心血一手建立的。在赴柏林交流的一年时间里，我的关注点很快落到了柏林联建住宅上。德方导师 Philipp Misselwitz 教授向我推荐了很多值得研究的案例，并提供了大量一手资料，它们大部分都出现在了此书中。在完成对柏林联建住宅项目的调研与走访后，我带着大量的资料回国并开始了相关的整理和深入研究工作。

在论文写作过程中，我一度不知从何着手。通过反复的讨论，李老师最终帮助我确定了论文的研究框架，指明了研究方向，这才使得接下来的论文写作得以顺利进行。

2014 年 6 月在我顺利完成毕业论文答辩之后，李振宇教授提议与我一起在现有论文的基础上，将研究成果整理出书，这既是对我研究生学习阶段的总结，又是作为学院该项目的首个双语出版学术成果，其意义可见一斑。随后我们向 Misselwitz 教授表达了这一想法，他也欣然答应加入到这份有意义的工作中来。

在接下来的一年里，我们又反复讨论了此书的主体结构以及表现形式，最终由我执笔，两位教授配合补充案例图文资料以及修改订正的方式合作将本书完成。在这一过程中，李老师还联系了王梓笛帮忙校对本书的英文稿，确保了本书

This book is revised based on my master dissertation of dual degree program between Tongji University and TU Berlin. During one year after the thesis defense, I supplement and improve the dissertation according to the suggestions from my supervisors. This book is also the first published bilingual research work of this program. I wish this book can open the subject for discussion and more joint graduate program researches can be shared with more people.

Since 2006 I majored in architecture in Tongji University. Influenced by the Bauhaus teaching model and international vision, I was fascinated by German architecture. During the summer school in Bern in 2009 I got acquainted with Prof. Li Zhenyu who become my supervisor later. It is Prof.Li lead me to the world of housing research. Without his support and encouragement, this book is hard to be published.

In 2012 I was chosen to join the urban design master dual degree program between Tongji University and TU Berlin, which is one of the earliest dual degree programs in our college and is established majorly by Prof.Li. During the year in Berlin, my attention was soon drew in Co-housing. My supervisor Prof. Philipp Misselwitz in TU Berlin recommended me a lot of cases worthy of research and offered me valuable materials, most of which appear in my book. After finishing the field trip of Co-housing in Berlin, I carried amounts of research material and got back to China. Then I started my research work in further.

After several times of discussion, Prof. Li helped me define the research frame of my dissertation and indicate the research direction, which ensure the writing go off without a hitch.

After my final thesis defense in June 2014, Prof. Li proposed that we should work together to publish my research work. On one hand it is the summary of my master degree, on

的顺利完成。

临近出版之际，首先要感谢我的导师李振宇教授，是他在我迷茫困惑的时候为我拨云见日，指明方向，并悉心教导我做研究的方式方法。长时间的耳濡目染使得整本书无论是写作逻辑还是研究架构都深受李老师的影响。同时李老师还为我提供了很好的出版平台，让我能够更加专注地完成本书的写作。

其次要感谢我的德方导师同时也是柏林工大联培项目的德方负责人 Philipp Misselwitz 教授，虽然相隔两地，但他仍能从百忙中抽空帮助此书的完成。

感谢王梓笛对本书英文稿的校对！感谢徐纺老师以及滕云飞，朱笑黎三位编辑在出版过程中给予的专业帮助！

最后，感谢谢路昕、感谢我的家人在背后默默的鼓励与支持，是他们以无私的爱给予我全方位的支持，使我能够心无旁骛地投身于研究工作，希望本书成为对他们最好的回馈！

龚喆
2015 年 12 月

the other hand it can be the first published bilingual work of our dual degree program. I accepted the invitation without hesitation and soon Prof. Misselwitz also joined us to complete this meaningful work. In the following year we discussed the main body and the outlook of this book and finally we finished the book.

With the publication drawing on, I shall show my gratefulness to my supervisor Prof. Li Zhenyu firstly. He taught me how to do research and how to solve problems logically.

Secondly I shall give my thankfulness to my German supervisor Prof. Philipp Misselwitz. He helped finishing a meaningful work to us.

Then I shall thank Wang zidi for her revision of my book's English version.

Thank Xu Fang, Teng Yunfei and Zhu Xiaoli for their help in publishing this book.

Finally I owe my gratefulness to my love Rowena and to my family. Every time I need them, they are just right behind me and support me selflessly. Wish this book be the best gift to them. Thank you and love you for ever!

GONG Zhe
December 2015